The **CEO**
who mocked

AI

(until it made him MILLIONS)

An AI Playbook Disguised as a Story - A Business Leader's
Guide to AI-Driven Transformation

AAMIR QUTUB

Published by Dumb Monkey Productions

ISBN (Paperback): 978-1-7640014-0-3

First Edition, 2025

Disclaimer:

This book is for informational purposes only. It is not intended as professional, legal, or financial advice. The author and publisher disclaim any liability arising from the use or misuse of the information contained herein.

Trademark Notice:

All trademarks, product names, and company names mentioned in this book are the property of their respective owners and are used for identification purposes only.

In memory of my mother, who fought bravely against cancer for three years. I hope for the day when AI finally finds the cure and brings hope to others.

Acknowledgments

Writing this book was far from a solo effort—it was powered by caffeine, deadlines, and the patience of some incredible people.

To my family—thank you for your love, understanding, and for tolerating my endless rants about AI, digital disruption, and the future of work.

To my team at Enterprise Monkey—you are the real heroes, keeping things moving while I disappeared into "writer mode." Your work and ideas shaped so much of what's in these pages.

To my mentors, peers, and the brilliant business minds I've had the privilege to work alongside—you've challenged me, guided me, and occasionally saved me from my own Dumb Monkey moments.

And finally, To Selene. Not the one behind the curtain—she's fictional (...or is she?). But to all the unseen thinkers, builders, and innovators shaping the future of AI and business—this book is as much yours as it is mine.

To you—the reader—thank you for trusting me with your time. If this book helps you stay ahead, avoid Dumb Monkey Syndrome, and maybe even laugh along the way— then we've both won.

Join the tribe—get exclusive resources, tools, and updates on all things AI, business, and avoiding Dumb Monkey Syndrome.

Visit **mnky.au/join** or scan the QR code.

Before You Begin – A Small Favour

I've created this book to help business leaders like you harness the power of AI—without the jargon.

If this book helps you in any way, I would be incredibly grateful if you could leave a quick review on Amazon.

Your feedback will not only help other leaders discover the book, but also encourage me to keep writing and creating more content that simplifies technology for businesses.

Leaving a review is quick and easy:

Scan the QR code or go to this link –

mnky.au/review

Share your honest thoughts—just a few lines would be amazing!

Thank you so much. Your support means the world to me.

Warm regards,

Aamir Qutub

Contents

Chapter 1

Scepticism vs Desperation

Blake Harrington, CEO of Harrington Construction, had seen bad quarters before— hell, he had even survived COVID. But this was different. Three consecutive losses. Stalled projects. Competitors swooping in, underbidding him at every turn. This wasn't just a rough patch—it was a crisis.

Blake sat at the head of the long mahogany conference table, arms crossed, jaw set like concrete. Across from him, his CFO, Katherine, clicked to the next slide in the quarterly report.

And suddenly, the room felt a lot smaller. Too much red. Profits were down 24%, bid success rates had plummeted, and worst of all? Steelworks Developments—a construction firm one-third his size—had outbid him. Again.

Blake exhaled slowly, tapping his fingers against the table. "Alright," he said. "Let's have it."

Katherine hesitated. "Blake, it's not just pricing. Steelworks is moving faster, winning more bids, and completing projects ahead of schedule."

Blake frowned. "That doesn't make sense. They're a fraction of our size. How the hell are they moving faster than us?"

Katherine shifted in her chair. "Well... they've undergone a massive digital transformation, powered by AI."

Blake blinked. Then he laughed. "Wait, wait, wait. You're telling me that a construction company—not a tech firm, not bloody NASA—is using AI?"

Katherine nodded. Blake looked around the room. Silence. Nobody in the room looked like they fully understood what was happening either. "...Alright," he said, rubbing his temples. "What does that mean? Are they using that, uh... chat thing?"

"ChatGPT?" someone murmured.

"Yeah, that. Or that pilot thing Microsoft was pushing?"

"Copilot."

Blake waved a hand. "Whatever. Point is, what does 'using AI everywhere' actually mean?"

Katherine sighed. "That's just it. I don't fully know. But every conversation in the industry keeps pointing to one thing: they have AI agents handling everything."

Blake stared at her. "...Everything?"

She nodded. From the other end of the table, his Head of Operations spoke up. "I heard they have AI handling bidding, procurement, scheduling, finance—hell, even compliance."

Blake's jaw tightened. "Okay, let's say that's true," he said, leaning forward. "Why don't we have this?"

Another silence. Then his Head of Strategy coughed. "Honestly, Blake? I don't know where to start."

Blake clenched his fists. This was the problem. There was too much damn information, and none of it told him what to do. Every time he searched online for AI, he got: Tech jargon-filled articles written for developers. Startup bros on LinkedIn saying "AI will change everything." Hyped-up success stories of companies that "transformed" overnight. And on the other hand?

His mates' companies—good, established businesses—were shutting down. It wasn't just Steelworks growing.

Blake exhaled sharply. Three quarters. Three losses. A record-breaking streak, but for all the wrong reasons. His company had weathered COVID, government shutdowns, supply chain nightmares. Yet somehow, this AI-powered shift was doing what a global pandemic couldn't—bringing him to his knees.

He needed answers.

That night, after a frustrating deep dive into AI rabbit holes, Blake rubbed his temples. The numbers didn't lie—his company was on track for its worst year since he'd started it. He had always been the guy who figured things out, who powered through crises. But this wasn't a supply chain issue or a financial crunch. This was a whole new world.

Finally, he picked up the phone. He dialled John Canning, the CEO of the State Business Chamber. If anyone knew what was happening in business, it was John.

"Blake! To what do I owe the pleasure?"

"John, tell me something." Blake took a deep breath. "Do you know what the hell is happening with AI?"

There was a pause. Then John sighed. "Mate... if I did, I'd be sleeping better at night."

Blake frowned. "Wait, you don't know either?"

"Not fully. I mean, I get the big picture, but the actual 'how do we use this in business' part? No clue."

Blake exhaled. "Great. So we're all blind."

"Pretty much," John chuckled. "That's why I started seeing someone."

Blake blinked. "Seeing someone?"

"Dr Monroe"

Blake frowned. "Wait, you're seeing a psychologist?"

John chuckled. "Not just any psychologist. Dr. Selene Monroe."

"Never heard of her."

"She's... different. Master's in psychology and a double PhD in Business Strategy and Digital Transformation. Worked with some of the biggest corporations in the world. Specialises in guiding business leaders through transitions."

Blake narrowed his eyes. "What kind of transitions?"

"The kind that make or break a company. Digital disruption, AI, automation, leadership reinvention."

Blake scoffed. "So, what? She just sits there and tells CEOs to 'embrace change'?"

John laughed. "No, mate. She makes you see what you're missing. When I talk to her, it's like she already understands what I'm struggling with—better than I do. She breaks it down so I can actually do something about it."

Blake leaned back, arms crossed. This was starting to sound like corporate voodoo. John continued. "She doesn't advertise. She only takes referrals. And there's a long waiting list... but I can get you in."

Blake hesitated. He hated this. But he hated losing even more.

"Alright," he muttered. "Set it up."

From: John Canning

To: Blake Harrington

Subject: Got You In – And Something Else

Blake,

I've pulled some strings—consider it sorted. You're in.

By the way, I've been following this guy on LinkedIn—Aamir Qutub. Honestly, bit of a nutcase. Keeps going on about how AI is going to change the way we work and disrupt every industry.

Half the time, he sounds like he's had too much coffee; the other half... he actually makes some bloody good points. Might be worth keeping an eye on him. Here's his profile - https://mnky.au/aqlinkedin

Oh, and apparently, he's writing some kind of business fiction book about AI and CEOs. I mean, who's reading that?

Talk soon,

John

Chapter 2

The AI Therapist

Blake hesitated outside the door.

He wasn't sure what annoyed him more—the fact that he was here, or the fact that John had made this sound like some elite underground club for business leaders who had 'figured it all out.'

Therapy. For Business Transformation. What a joke.

And yet... He exhaled and pushed the door open. The atmosphere shifted immediately. It was subtle, but undeniable. The room was dimly lit, yet warm, with a glow that felt almost intimate—like a high-end whiskey lounge or a place where secrets were whispered over candlelight.

And then, there was the scent. Soft, but distinct. Vanilla, sandalwood, and a hint of jasmine. It wasn't overpowering. It lingered—just enough to draw you in.

Blake tensed. He wasn't used to spaces like this. He wasn't used to feeling like he'd stepped into a place meant to disarm him. And then his eyes settled on the curtain. Deep navy blue, stretching from floor to ceiling. It should have felt ridiculous, talking to someone hidden behind a fabric wall. But something about it made her presence more intense.

He could almost make out a shadow on the other side—a silhouette that hinted at movement, a shift in posture, the tilt of a head. His mind filled in the gaps. The way she might be sitting— poised, elegant, yet effortlessly confident. The way she might be watching him right now, amused at his discomfort. It was strange. He knew he couldn't see her. And yet, he could feel her.

"Mr. Harrington." Blake froze. That voice. Smooth, low, effortlessly composed. There was a warmth to it, but also an edge—a quiet control. The kind of voice that drew you in before you even realised it was happening.

A flicker of something unexpected passed through him. He swallowed. That was not what he had expected.

"Please, settle down on the lounge chair."

He hesitated for a second, then lowered himself onto the plush, oversized leather chair—the kind therapists probably ordered in bulk to make people feel vulnerable. It was annoyingly comfortable. "Alright, Doctor," he muttered. "Let's get this over with."

A soft chuckle from behind the curtain. "Call me Selene."

Blake frowned. "What?"

"That's what I prefer," she said, voice smooth, teasing. "Doctor feels so... clinical, don't you think?"

Blake shifted in his seat. Somehow, Selene felt even more mysterious than Doctor Monroe. Dangerous, even. But he wasn't about to admit that. "Fine," he muttered. "Let's get this over with, Selene."

"Let me guess—you think this is all a waste of time?"

Blake smirked. "Honestly? Yeah. I don't see how sitting here, talking about AI, is going to fix my business."

A pause. Then, a soft chuckle. "Who said anything about AI?"

Blake's brow furrowed. "Wait, isn't that—?"

"Let's talk about what's actually broken."

Blake exhaled sharply. "You mean besides the fact that my damn competitors—who are half my size—are winning while I'm stuck in the mud?"

Selene's voice remained calm, controlled. "Sounds frustrating."

Blake leaned forward. "Frustrating? Try insane. These guys don't have my reputation, my resources, my connections. Yet somehow, they're underbidding me, outpacing me, and finishing projects in record time."

A brief silence. Then, Selene's voice carried a playful lilt. "So tell me, Blake—why do you think that is?"

Blake ran a hand over his jaw. "I don't know. Maybe they're cutting corners."

"Is that what you really think?"

Blake hesitated. He had run the numbers, asked around. No red flags. If anything, their operations seemed smoother, more efficient. Which was even more frustrating.

"Then tell me this," Selene continued. "If it's not cutting corners, then what's changed?"

Blake sighed. "The industry's moving faster. Tech. Automation. All that stuff."

Selene's voice was calm but pointed. "And where does that leave you?" Blake frowned. Then Selene delivered it, perfectly timed. "Tell me, Blake—how do you feel about the AI you already use?"

Blake scoffed, leaning back in his chair. "I don't use AI." He shook his head. "In fact, I don't even allow anyone on my team to use that chat thing."

A soft chuckle. "That's cute. You're wrong, but cute."

His eyes narrowed. "Oh yeah? And how exactly am I wrong?"

A pause, deliberate. Just enough to let the question linger. "AI is already making decisions for you, whether you realise it or not."

Blake snorted. "No, it's not."

A slight shift, the sound of fabric brushing against itself. "Tell me, Blake—did you use Google Maps to get here?"

"...Yeah, so?"

"That's AI predicting traffic patterns in real-time."

Blake waved a hand. "Yeah, but that's just a map."

"Did you check your emails this morning?"

"...Obviously."

"Did your inbox filter out spam?"

Blake paused. "...Yeah?"

"That's AI deciding what you need to see and what's junk."

Blake shifted. She kept going. "Did you browse LinkedIn or read any news articles today? Did you get a notification about a 'recommended article'? Does your bank flag suspicious transactions?"

Blake stayed silent. Selene let it sink in before delivering the kicker. "Even your competitors' pricing strategies—they're not guessing, Blake. Their AI models are adjusting bids in real-time, predicting costs before you even submit yours."

Blake's stomach twisted. Because that—that actually made sense. A long pause. Then Selene said, almost casually, "Have you thought about what happened to Mick Dawson's business?"

Blake stiffened. Mick - A good mate, great businessman. Built Dawson Civil & Concrete from scratch. One of the best in the game. And then, one day—gone. Liquidated. Wiped off the map. Blake shifted in his seat. "Yeah... I heard."

Selene's voice was calm, measured. "You ever wonder *why*?"

Blake exhaled. "I don't know. Margins got tight. Market changed."

Selene hummed softly. "That's what he told you, isn't it?" Blake didn't respond. "Tell me," she continued, "do you think Mick's business failed because of bad luck... or because of Dumb Monkey Syndrome?"

Blake blinked. "...I'm sorry, what?"

Blake scoffed. "Oh, for—another bloody syndrome?" He leaned back, arms crossed. "First, it was COVID. Then inflation. Now we've got Dumb Monkey Syndrome? What's next—AI Anxiety Disorder? Should I start a support group?"

Selene laughed—genuinely. "Blake, relax. It's not that kind of syndrome."

Blake leaned back, still sceptical. "Right. So, what is it then? And more importantly—do I have it?"

Selene's voice turned playful but firm. "Ah, now we're asking the right questions."

Blake crossed his arms. "Alright, tell me."

Selene stayed silent for a beat. Then: "No. Not yet."

Blake blinked. "What do you mean, *not yet*?"

"You first need to decide—do you even want to continue this journey?"

Blake groaned. "Journey? What is this, a self-help retreat?"

Selene chuckled. "I get it, Blake. You're impatient. You want quick answers. But here's the thing—you didn't build your business in a day, right? You figured things out step by step." Blake didn't reply. She continued, "So, if you want to understand what's happening in business today, you have to start the journey. And that means doing your homework first."

Blake raised an eyebrow. "Homework?"

"Go home," Selene said smoothly.

Blake rolled his eyes. "Brilliant advice. I could've gotten that from my dog."

She ignored him. "Go home, create a free ChatGPT account, and start talking to it."

Blake squinted. "What the hell am I supposed to talk about?"

"Anything. Ask it to draft an email. Ask it how to negotiate a contract. Ask it to tell you a joke. Just start. See how it responds."

Blake exhaled. "That's it? Just chat with a bloody chatbot?"

"Yes," she said patiently. "Just talk to it. Play with it. See what it can do."

Blake rubbed his temples. He had come in expecting some big strategy session. Instead, he was being told to go home and talk to a damn robot. But deep down, he knew something was shifting. Because for the first time in months, he wasn't just frustrated. He was curious. Blake stood up to leave. As he reached the door, he turned back.

"Alright, Doctor," he said, forcing a smirk. "Next time, you're gonna tell me about this Dumb Monkey Syndrome, right?"

Selene's voice was smooth, playful, but firm. "Only if you do your homework."

Blake groaned, stepping out. He already regretted this. And yet, later that night, when he hesitantly typed "ChatGPT" into his browser...

The screen loaded. A blank chatbox. A blinking cursor.

He hesitated. Then, finally, he started typing. One question. Then another. And another.

His fingers moved faster. His frown deepened. Until, finally— He leaned back. Exhaled. Stared at the screen. And muttered under his breath:

"What a load of crap."

From: Dr Selene Monroe

To: Mr Blake Harrington

Subject: Our Discussion – Exploring Further

Mr Harrington,

It was a pleasure speaking with you today.

As we discussed, ChatGPT is a useful starting point, but it is merely one option. There are other AI models—Gemini, Claude, and several more.

Each has its own strengths and nuances. Much like selecting the right blend for your morning coffee, the best way to understand them is to experience each for yourself.

I am sharing a resource that provides a clear comparison. You may find it helpful as you begin to explore: https://mnky.au/llm

Take your time. Experiment. See what resonates.

Until next time,

Dr Selene Monroe

Chapter 3

Dumb Monkey meets AI

Blake Returns – Still Sceptical. He pushed open the door, feeling a mix of impatience and reluctant curiosity.

"I'm back," he announced, dropping into the chair. "Now tell me about this Dumb Monkey Syndrome."

Selene chuckled softly from behind the curtain. "Did you do your homework?"

Blake exhaled sharply. "Yeah. I did. But honestly?"

He shrugged. "This ChatGPT thing is fine. Fun, sure. But it's too… generic. It goes off on random tangents. It's not fit for real business use."

Selene tilted her head. "Interesting. What exactly did you ask?"

Blake cleared his throat, pulling out his phone. "Alright, let's see... I searched: *Business improvement for a construction business , Welcome email for customers , Best way to increase profit*" He looked up. "And half the time, it just spat out some generic blog post-sounding answer."

Selene went completely silent. Then she laughed—hard.

Blake's eyebrow twitched. "What?"

Selene tried to contain herself. "Oh, Blake..." she sighed. "I asked you to talk to it. Not to Google search."

Blake stiffened. "What's the difference?"

Selene leaned forward slightly. "Tell me, if you walked into a meeting with a business expert—say, a CEO who scaled five companies—would you just blurt out 'Business improvement for construction'?"

Blake hesitated. "No... I'd explain my situation first."

Selene nodded. "Exactly. So why would you expect AI to read your mind?"

Blake rubbed his temple. "Alright. So how should I have asked?"

Selene smiled. "Let me show you. Let's take your 'Business improvement for a construction business' search. That's like walking into a consultant's office and just saying, 'Make my business better.' It's vague. There's no direction.

Selene leaned in, watching Blake carefully. "Alright, let's rewrite that properly. You need to give AI the right job description before you expect the right answer."

She began typing:

Prompt for business improvement strategies:

You are an award-winning business strategist with years of experience in commercial construction. You have studied and analysed the most successful case studies worldwide, and you will think at your maximum efficiency to provide the best possible answer.

Before responding, you will evaluate your own answer and refine it further to ensure it aligns with industry best practices, cutting-edge technology, and my specific business challenges.

My situation:

I run a mid-sized commercial construction company in Australia with 70 employees. We focus on office buildings and industrial projects.

The challenge:

Recently, we have been losing bids to smaller competitors who seem to be delivering projects faster and at lower costs. I suspect they are using AI or digital tools that we have not adopted yet.

Key problems to solve:

Project delays due to material shortages and supply chain inefficiencies.

High overhead costs, particularly in labour and administration.

Losing bids because we cannot price as aggressively as competitors who leverage AI.

What I need from you:

Suggest three highly effective strategies (preferably with AI-driven or digital solutions) that have successfully improved business performance for companies like mine.

Provide real-world case studies or examples where these strategies have been implemented successfully.

Ensure your response is structured, data-backed, and provides actionable recommendations.

At the end of your response, evaluate your own suggestions and refine them to make sure they are the most practical and high-impact solutions for my use case.

Blake scratched his chin. "Alright, explain why this version is so much better."

Selene leaned back, her tone confident. "Because now AI is not just throwing you a basic response. It is being forced to think strategically, just like a real consultant would."

"Break it down for me," Blake said.

You are giving AI an identity

"You are an award-winning business strategist..."

This frames AI's role, making it more likely to give high-quality, expert-level insights. Without this, AI might respond like a generic blog post instead of a top-tier consultant.

You are pushing AI to analyse and refine its own response

"Evaluate your own answer and refine it further..."

Most people take AI's first response as final, but AI can analyse and improve its own work when prompted. This forces AI to check for flaws before giving a final answer.

You are providing deep context

"I run a mid-sized commercial construction company..."

AI now knows your industry, business size, and location, so its response will be more relevant. Without this, AI gives generic business advice that might not apply to construction.

You are telling AI what problems to solve

"Project delays, high costs, losing bids..."

Instead of letting AI guess what you need, you tell it exactly what challenges to focus on. This removes guesswork and makes the answer laser-focused.

You are structuring the output

"Suggest three highly effective strategies..."

AI can wander if not directed. By specifying three, you get a clear, structured response. Also, you are asking for real-world case studies, so AI prioritises examples over vague theory.

Blake exhaled. "Damn. So this whole time, I was not using AI wrong—I was just giving it vague instructions?"

Selene smirked. "Exactly. AI is just a tool. If you do not know how to use it, it is not AI's fault."

Blake sighed. "Alright, what about my welcome email for customers?"

Selene smirked. "Did you just type in 'Welcome email for customers' and expect poetry?"

Blake grumbled. "I may have."

Selene chuckled. "Well, that is why it gave you something generic. AI does not know who your customers are unless you tell it."

Blake leaned forward. "So I have to be specific about my business?"

"Exactly," Selene said. "Who you are, what you offer, what the email should do. It is like briefing a copywriter—details matter."

She pulled up a revised version of the prompt.

Prompt for Welcome Email:

You are an experienced marketing strategist with deep expertise in customer engagement and retention. You have worked with some of the most successful companies in the world, crafting high-converting onboarding strategies that improve client satisfaction and long-term loyalty. Your success rate in writing effective customer emails is unmatched, and you bring years of knowledge on what makes an introduction email truly impactful.

Before finalising, you will evaluate your response and refine it to ensure it is engaging, clear, and aligned with best practices.

My situation:

I own a commercial construction company that specialises in mid-sized office buildings. We have just launched a new customer onboarding process where we offer clients a dedicated project manager and a real-time project tracking dashboard.

What I need from you:

Write a professional yet warm welcome email that does three things:

Thanks the client for choosing us.
Briefly explains what happens next in the onboarding process.

Introduces their dedicated project manager and the dashboard.

Keep it concise, about 200 words, ensuring it is informative but engaging. Use a tone that is professional, clear, and welcoming, without being overly casual or robotic.

Once written, review the email and refine it to make sure it is impactful and easy to read.

Blake frowned. "So, what exactly makes this version better?"

Selene smiled. "Because now AI actually knows what you want. Let's break it down."

You are giving AI a specialised role with expertise

"You are an experienced marketing strategist with deep expertise in customer engagement and retention..."

This forces AI to take on a specialised approach rather than acting like a generic text generator. By referencing past work with successful companies, AI draws from high-quality knowledge rather than giving surface-level responses.

You are telling AI to refine its own answer before presenting it

"Before finalising, evaluate your response and refine it..."

AI can generate multiple versions of the same prompt. By telling it to refine itself, you ensure the best possible version is given first. This prevents vague or repetitive responses.

You are giving AI real business context

"I own a commercial construction company specialising in mid-sized office buildings..."

AI now understands who your customers are, your industry, and your unique selling points. Without this, AI would generate a generic welcome email that does not reflect your business.

You are structuring the email's content clearly

"Write a professional yet warm email that does three things..."

This gives AI a clear structure, so the output is focused and actionable. Without this, AI might create an unfocused or wordy email that does not hit the right tone.

You are setting constraints to ensure clarity

"Keep it concise, about 200 words..."

If you do not set a word limit, AI might generate something too long or too short. This ensures the right balance between detail and brevity.

Blake exhaled, rubbing his forehead. "So, this whole time, AI was not the problem—I was just giving it half-baked instructions?"

Selene smirked. "Exactly. AI is only as good as the prompt you give it. The better your input, the better your output."

Blake leaned back, nodding. "Yeah... I see why my version did not work."

Selene smiled. "Good. Now, let's see if you actually put this into practice."

Blake waved a hand. "Alright, last one. I asked how to increase profit."

Selene let out a dramatic sigh. "And what did you expect? A three-step shortcut to instant millions?"

Blake shrugged. "Would have been nice."

Selene shook her head, amused. "Alright, let's do this properly." She pulled up a new version of the prompt.

Prompt for increasing profitability:

You are a highly sought-after financial strategist who has advised Fortune 500 companies on profitability, cost optimisation, and business efficiency. You have successfully helped businesses increase their net margins by leveraging advanced pricing strategies, operational improvements, and AI-driven financial insights.

Before finalising your response, you will analyse your own answer, refine it for clarity and practicality, and ensure that it aligns with proven business strategies that have led to measurable financial growth.

My situation:

I run a commercial construction business with an annual revenue of $15 million, but our profit margins have been shrinking. Right now, our net profit is at 8 percent, and I want to get it to at least 12 percent.

Challenges affecting profitability:

Rising material costs due to unpredictable supply chain fluctuations.

High payroll expenses that are eating into our margins.

Clients pushing for lower pricing, reducing our ability to remain competitive.

What I need from you:

Provide three high-impact strategies for increasing net profit while maintaining quality and efficiency.

Break the strategies into short-term (next 3 months) and long-term (1 year and beyond) so I can implement them in phases.

Offer specific AI tools, automation solutions, or industry best practices that have been proven to improve profitability in construction or similar industries.

Once complete, refine your answer to make sure it is practical, actionable, and tailored to my use case."

Blake exhaled, shaking his head. "So what you are saying is... AI is not useless. I was just asking it the way an intern fresh out of college would."

Selene grinned. "Pretty much. You were using a Ferrari like a bicycle."

Blake ran a hand through his hair. "Alright, fine. This actually makes sense. But tell me one thing—if AI is so smart, why does it not just tell me to structure my question better in the first place?"

Selene smirked. "That is the difference between AI and humans. AI gives you what you ask for. A good consultant? They tell you when you are asking the wrong question."

Blake leaned back, nodding slowly. "Yeah... I need to rethink how I use this thing."

Selene smiled. "Now you're getting it."

Blake tilted his head. "Alright, so is there an actual framework for this?"

Selene sighed dramatically. "You business nerds need a framework for everything, don't you?"

Blake smirked. "Well, yeah. It helps."

Selene chuckled. "Alright, I'll give you some frameworks... once you do your homework properly."

Blake sighed. "Fine. Points for effort, at least?"

Selene's voice was light but approving. "I'll give you points for trying."

She let a pause hang in the air, then added, "Now, let me tell you about Dumb Monkey Syndrome."

Selene leaned in slightly, her voice taking on a storytelling rhythm. "Evolution doesn't wait. Apes evolved. Humans evolved. Businesses? Same rule applies."

Blake raised an eyebrow. "Is this a biology lesson, or are we getting to the part where I make money?"

Selene smirked. "Patience, CEO. If humans had stopped at the wheel, we'd still be pushing carts. The same way companies that stop evolving get left behind."

Blake exhaled. "So you're saying companies that refuse to innovate... go extinct?"

"Exactly."

Blake leaned forward.

"That's Dumb Monkey Syndrome."

She let the words sink in before adding,

"People or businesses that fail to evolve with time eventually die."

Blake was silent. Then, Selene delivered the gut punch.

"You've seen it before. Kodak. Nokia. Blockbuster."

Blake exhaled. "Yeah."

"And the same thing," she continued, "happened to Mick."

Blake shifted uncomfortably.

"Look at his competitors," Selene continued. "They evolved. They digitised. They streamlined processes, adopted new tools, AI-driven cost estimations, automated workflows."

Blake ran a hand through his hair.

"And Mick?" Selene asked. "He stuck to spreadsheets and old-school practices. He told himself, 'If it ain't broke, don't fix it.' He kept cracking coconuts."

A beat.

"But the industry changed," she continued. "And Mick? He didn't."

Blake exhaled sharply. "...Damn."

Selene's voice softened, but didn't lose its edge.

"Blake, I don't want you to be Mick."

A long pause.

Blake looked at the curtain, as if he could see through it.

He suddenly felt like he was standing at a crossroads.

That night, Blake opened ChatGPT again.

He stared at the blank input box.

Then he typed differently.

And this time?

The answer made him sit up straight.

And for the first time, AI wasn't just a buzzword. It was a tool. And maybe—just maybe—it could actually work.

From: Dr Selene Monroe

To: Blake Harrington

Subject: A Starting Point – Prompts for Business Leaders

Mr Harrington,

Following our recent discussion, I thought you might find this helpful.

I am sharing a curated list of 50 handpicked prompts specifically designed for business leaders exploring AI tools like ChatGPT and others.

These are not generic. Each is crafted to help you get practical, business-relevant insights—whether it's refining your bidding strategy, optimising project schedules, or even drafting that next difficult email.

You will likely develop your own style over time, but consider this a starting point:

https://mnky.au/aiprompts

As always, it's not just what you ask—but how you ask.

Until next time,

Dr Selene Monroe

Chapter 4

Smell of Selene & Frameworks

Blake pushed open the door, stepping into the now-familiar space. And just like last time, the moment he entered, the scent hit him first.

But it was... different. Still that same soft vanilla and sandalwood base—but tonight, there was an added layer. Something warmer, richer. Amber? A hint of spice? Maybe even a touch of musk?

It was still Selene's perfume—that much he was sure of. But she'd changed it. Subtly. Deliberately.

Blake wasn't into women's perfumes. He couldn't tell a Chanel from a Dior. But this? This, he noticed. And the worst part? He'd done the same damn thing.

He adjusted his cuff, inhaling the slight trace of Tom Ford Oud Wood he'd applied before leaving. Not his everyday cologne. His favourite one. Reserved for special occasions. Not that this was one. Obviously.

He shook off the thought, reminding himself that he still hadn't even seen the woman. And yet, his mind insisted on picturing her. Was she casual, effortlessly elegant, sitting cross-legged with a teasing smirk? Or was she the polished, powerful type, posture straight, one hand resting lightly on her lap as she analysed him like an open book?

He clicked his tongue. This is a business meeting, not a bloody date. Blake walked in, forcing his thoughts back to why he was here. The curtain, as always, remained closed. And, as always, Selene's voice greeted him before he could speak.

"Back again so soon, Blake?"

He settled into the plush lounge chair, exhaling.

"I hate to admit it... but that ChatGPT thing—it's actually kind of impressive."

Selene's voice carried a quiet satisfaction. "Ah. The reluctant compliment. My favourite."

Blake smirked. "Don't get used to it."

A soft chuckle. "So tell me—what changed?"

Blake rubbed the back of his neck. "Well, I actually tried talking to it. Properly, like you said."

"And?"

Blake leaned forward slightly. "And... it gave me a detailed, step-by-step plan for improving my bidding process. Like, actual strategies that made sense."

Selene nodded approvingly. "Go on."

Blake exhaled, still processing. "I even asked it to rewrite an email for me. The damn thing sounded better than I do."

Selene grinned. "That's a low bar, Blake."

Blake pointed at the curtain. "Alright, alright. No need to get cocky, Doctor—"

Selene cut in smoothly. "Selene."

Blake smirked. "Right. Selene." There was a strange satisfaction in saying her name out loud.

As he settled back into the chair, his mind started to wander again. The perfume. The voice. The confidence. It was all designed to draw you in. Blake shook off the thought.

"This is progress," Selene said, her tone approving. "But AI, like any tool, is only as good as the person using it."

Blake raised a brow. "Are you saying I'm bad at using it?"

Selene paused for a second. Then, with mock sympathy—

"...Yes."

Blake rolled his eyes. "Fantastic. Another ego boost."

Selene chuckled. "You're improving. But you're still missing structure. You need a method. A system."

Blake folded his arms. "This isn't just an excuse to make me memorise more acronyms, is it?"

Selene chuckled. "No, Blake. This is about making AI work for you—like a high-performing employee, not a random search engine. Each of these methods applies to different business scenarios. If you use AI the right way, you'll get structured, actionable insights rather than vague, useless responses."

She leaned forward slightly.

"Think of it this way—each of these frameworks solves a specific kind of problem. The key is knowing when to use them and how to apply them properly."

1. The S.C.O.R.E Method – For Business Strategy & Decision-Making

You will use the S.C.O.R.E method when you need AI to assist in high-level decision-making, strategic planning, or solving complex business problems. This method is particularly useful when you're trying to optimise operations, improve efficiency, or expand into new markets but need structured insights rather than generic advice. If you're facing a critical challenge in your business and need a clear roadmap, S.C.O.R.E ensures AI gives you focused, detailed, and practical solutions.

How to Apply It

Blake raised an eyebrow. "Alright, so how does this thing work?"

Selene tapped the table. "S.C.O.R.E forces you to structure your question like you would for a consultant."

S – Situation: What is happening?

C – Challenge: What problem are you facing?

O – Objective: What is your desired outcome?

R – Request: What exactly do you need AI to do?

E – Expectation: How should the answer be structured?

Example Prompt Using S.C.O.R.E:

I run a mid-sized commercial construction firm specializing in office buildings and industrial projects. Over the last six months, we have experienced significant project delays due to material shortages and inefficient scheduling. As a result, our bid success rate has dropped, and we are struggling to remain competitive.

The biggest challenge we face is improving our scheduling and material procurement process to prevent delays.

Our objective is to reduce project delays by at least 20% over the next six months while maintaining cost efficiency.

Based on this, I want you to act as a business strategist and provide a three-step action plan that includes:

Specific AI-driven tools or digital solutions we can implement to optimise scheduling and procurement.

A case study or real-world example of a construction company that has successfully addressed this issue.

A risk assessment, highlighting potential obstacles and how to mitigate them.

Structure your response in a way that allows for easy implementation.

Blake nodded. "Alright, that actually makes sense. It's like giving AI a consulting brief instead of just throwing questions at it."

Selene smirked. "Exactly."

2. The T.E.A.C.H Method – For Getting AI to Explain Complex Topics Clearly

There will be times when you need to quickly grasp a complex concept—maybe AI in construction, blockchain in supply chains, or advanced financial modelling. The T.E.A.C.H method helps when you need to break down technical or industry-specific topics into simple, actionable insights. It's especially useful when you have to train your team, explain something to a client, or prepare for a meeting where you need to sound like you actually know what you're talking about.

How to Apply It

Blake snorted. "Alright, so this one is basically AI for dummies?"

Selene smirked. "More like AI for busy executives who need clear, no-fluff answers."

T – Target audience: Who is this explanation for?

E – Explanation style: Do you want it simple, technical, or analogy-based?

A – Application: How does this concept apply to your business?

C – Constraints: Should it be under 200 words? No jargon?

H – Hook: Make it engaging—use a metaphor, comparison, or story.

Example Prompt Using T.E.A.C.H:

I need to explain AI-driven predictive analytics to construction project managers who are experienced in site work but have no technical background in AI.

Assume they are unfamiliar with terms like algorithms or machine learning.

Explain it using a real-world construction analogy—something site-related that will resonate with them.

Keep it under 250 words, avoid technical jargon, and use a simple, conversational tone.

Finally, give one practical example of how predictive analytics could help reduce delays or cost overruns on a construction site— something they can immediately visualise.

3. The R.A.F.T Method – For Creating AI-Generated Content

There will be situations where you need to write emails, reports, or proposals quickly and professionally but do not have the time to fine-tune every word. The R.A.F.T method is designed to help AI create clear, structured, and engaging content tailored to your business needs.

How to Apply It

Blake smirked. "Let me guess—this one stops AI from writing emails that sound like a scam?"

"Exactly," Selene said. "You tell it who to be, who to talk to, and how to say it."

R – Role: Who is AI acting as? A copywriter? A business consultant?

A – Audience: Who is the message for?

F – Format: Is this an email? A proposal? A LinkedIn post?

T – Tone: Should it be formal, persuasive, engaging?

Example Prompt Using R.A.F.T:

I need you to act as a senior business consultant with expertise in the construction industry and AI-driven business communication.

Please write a professional, client-facing email draft announcing the rollout of our new AI-powered project tracking system. The system helps clients get real-time updates on their project status, detect potential delays early, and improve overall project transparency.

Our clients are construction developers and site managers—they value efficiency, clarity, and straight-to-the-point communication.

The tone should be formal but approachable—professional, yet easy to understand.

Keep it concise—no more than 250 words—while making the benefits of the system clear (e.g., faster updates, fewer surprises, cost savings).

Ensure the email closes with a clear call to action—inviting clients to book a demo or **reach out to their project manager for more details**.

4. The S.T.A.R Method – For AI-Driven Problem-Solving & Brainstorming

There will be times when you need creative, out-of-the-box solutions for your business—whether it is improving employee engagement, increasing customer satisfaction, or optimising project workflows. The S.T.A.R method is designed for situations where you do not want AI to simply regurgitate best practices but instead generate fresh, innovative solutions tailored to your specific needs.

This method is particularly useful when you are brainstorming new ideas, tackling unique challenges, or seeking alternative ways to optimise operations. Instead of generic advice, AI will provide targeted recommendations that are more insightful and actionable.

Blake leaned forward. "Alright, so this one's about making AI more creative?"

Selene nodded. "Think of AI like a brainstorming partner. If you don't give it direction, it'll just list generic ideas. The S.T.A.R method forces AI to generate solutions that are specific, practical, and aligned with your business goals."

S – Situation: What problem are you trying to solve?

T – Target outcome: What do you want to achieve?

A – Approach: Should AI suggest best practices, case studies, or experimental ideas?

R – Refinement: Ask AI to improve its answer and suggest the best option.

Example Prompt Using S.T.A.R:

I am the CEO of a mid-sized commercial construction firm. We have been struggling with on-site employee engagement, especially during safety briefings.

Currently, workers find the safety sessions repetitive, tune out during meetings, and do not retain critical safety protocols, leading to an increase in minor accidents and compliance issues.

My goal is to improve engagement and knowledge retention in safety training while ensuring compliance with industry regulations.

I want you to act as an HR and workplace behavioural specialist and provide me with three innovative approaches to make safety training more interactive and engaging. Focus on using behavioural psychology, gamification techniques, or real-world case studies where companies have successfully improved safety engagement.

Once you provide these solutions, refine your answer by selecting the best one based on practicality and ease of implementation for a construction workforce.

Blake stretched his arms. "Alright, I'll admit—this is actually useful. But tell me, are these the only four frameworks, or are you making me memorise the greatest hits?"

Selene chuckled. "Not even close. There are many different frameworks, each designed for different types of AI interactions. These four are a solid starting point, but they're just a guide. AI is like any tool—you'll get the best results when you develop your own way of using it."

Blake raised an eyebrow. "So, there's more?"

Selene leaned back. "Plenty. I'll send you an email with a full list of advanced AI frameworks that you can explore later. But don't overcomplicate it. The key is not to memorise them—it's to understand that structuring your input leads to smarter, more strategic AI responses."

Blake tapped his fingers on the chair's armrest. "So, I'll figure out what works best for me?"

Selene nodded. "Exactly. You're not just using AI—you're managing it."

Blake leaned back. "Alright. This is actually useful." He hesitated. "But there's one thing that still bothers me—how do I know AI won't steal my data?"

Selene nodded.

"Good. You should be thinking about that."

Blake raised an eyebrow. "Wait, did you just say I'm right about something?"

Selene laughed. "Don't get used to it."

Her voice turned more serious. "Look, not all AI tools are the same. Some store data, some don't. Some are built for privacy, some aren't. We'll go deeper into that later, but for now—be smart about what you input. AI is a tool, not a vault."

Blake frowned. "So, don't put anything confidential into it?"

"Exactly," Selene said. "And use AI tools that are built for business security—ones that let you control where data is stored and processed. AI is like a junior employee—helpful, fast, but you wouldn't trust it with your deepest secrets."

Blake nodded slowly. "Alright. Fair enough."

Selene leaned forward slightly. "Now, Blake, I want you to take one of these frameworks and use it for something that actually impacts your business."

Blake exhaled. "Homework again?"

Selene grinned. "You're catching on."

He took a few steps toward the door, then paused.

."By the way..." He hesitated for just a second. "Nice perfume."

A brief silence. Then—a small, knowing smile in her voice.

"Noted."

Blake exhaled, smirking slightly to himself as he pushed open the door.

From: Dr Selene Monroe

To: Mr Blake Harrington

Subject: Prompt Frameworks

Hi Blake

Good to see you today.

We discussed four core frameworks—S.C.O.R.E, T.E.A.C.H, R.A.F.T, and S.T.A.R.—to structure your AI prompts. But these are not the only ones. There are many more, each suited to different business needs.

I am sharing a list here:

https://mnky.au/promptframeworks

Explore at your pace. The right approach often unlocks the right answers.

Until next time,

Dr Selene Monroe

Chapter 5

Tender in trouble

Rumours spread fast. And the rumour right now? Blake's company was on the verge of collapse. People were jumping ship.

Three of his top salespeople had already quit—the same guys who handled big government tenders. They had all the institutional knowledge, all the contacts, and now... they were gone. The ones left? Overworked, uncertain, and already polishing their resumes.

And now, this government tender. A massive contract. Highly regulated, complex, and with enough paperwork to bury a small law firm. If they didn't win this, it would be bad. Really bad.

Blake did the only thing he knew how to do. He rolled up his sleeves. He cleared his schedule. And he dove in.

Blake hadn't worked this hard in years.

Meetings?----Cancelled.

Emails?---Ignored.

Calls?----Straight to voicemail.

His secretary, Katrina, rescheduled everything—twice. Then a third time. He barely noticed. He barely cared.

Because this tender? It was a beast. Hundreds of pages. Policies, compliance docs, financial statements, project histories. Every section had to be perfect. One mistake? Disqualification. And he was doing it alone.

By 9 PM, the office was dark, except for the glow of his laptop screen. His dinner sat untouched. His head pounded. And then— his phone rang.

Private Number. Blake frowned. Who the hell calls this late?

Only a handful of people had this number.

His kids. A couple of close friends. His ex-wife (who, thankfully, never called).

It rang once. He stared at it. Spam? A mistake?

It rang again. With a sigh, he picked up. "Blake speaking."

A brief silence. Then, that voice.

Smooth. Playful. A little too knowing. "Mr. Harrington."

Blake blinked. His exhaustion vanished—just for a second. He knew that voice. He leaned back in his chair, smirking. "Well, well... checking in? Or missing me?"

Selene laughed softly. "Your secretary has rescheduled your appointments three times. I was beginning to think you'd developed a deep-rooted fear of curtains."

Blake chuckled. "I've just been a little... busy."

Selene's voice softened. "You sound tired."

Blake exhaled. "That obvious?"

He rubbed his eyes, glancing at the mountain of paperwork. Then, without thinking, he started talking. Explained the situation. The tender. The departing employees. The impossible workload. Then he added, almost as an afterthought—

"And before you start, no, AI can't help with this. Some problems just have to be solved the hard way."

A pause. Then, very calmly—

"That's true... but this one can."

Blake frowned. "Come again?"

Selene's voice was steady, confident. "You need a system. A process. Something that organises, retrieves, and structures information instantly. Something that can generate responses, analyse compliance, and cross-check details in seconds."

Blake rubbed his temple. "That sounds great, Doc. But unless you've got an army of AI robots—"

Selene's voice was calm, assured. "You don't need an army."

Blake frowned. "Then what?"

"You just need one. A custom GPT—built specifically for tender preparation."

Blake sat up straighter. "Custom... GPT?"

"Yes. Inside ChatGPT, you can actually build your own version—fine-tuned for your needs. Instead of asking generic questions every time, you can train it with your business data, your style, your requirements."

Blake was silent for a moment. Processing.

Selene continued, "It won't do everything for you—but it can cut your workload by 80%. Organising data, summarising compliance requirements, drafting responses—it's like having a personal AI assistant for tenders."

Blake leaned back, sceptical but intrigued. "You really think this can work?"

"I know it can."

"Set a Zoom call tomorrow morning. I'll walk you through it on screen share."

A pause. Then, softly—just enough to make him wonder if he imagined it—

"Goodnight, Blake."

He sighed, ready to hang up—then, just as he moved his thumb to end the call—

"And Blake?"

He hesitated. "Yeah?"

"Make sure you get some sleep."

Click.

Blake stared at the phone for a second. Then at the tender documents. He should sleep. But not yet.

From: Dr Selene Monroe

To: Blake Harrington

Subject: Custom GPT – Getting Started

Hi Blake,

Hope you managed to get some sleep.

Here's a guide on how to build your own Custom GPT:
https://mnky.au/customgpt

I'll walk you through it in detail when we meet, but feel free to explore. It's easier than you think.

Until next time,

Dr Selene Monroe

Chapter 6

Custom GPT Saves the Day

Blake stared at the Zoom link. A deep sigh. This was not how he had imagined his morning.

If someone had told him a year ago that he'd be scheduling Zoom calls with an unseen AI expert-slash-therapist to build some kind of robotic tender writer, he would have laughed in their face.

Now?

He hesitated for a second before clicking.

The screen loaded. The familiar curtain of mystery remained— Selene's camera was off. Of course.

But she was there.

"Right on time, Mr. Harrington."

Blake took a sip of his coffee. "You sound surprised."

"Not at all. I just assumed you'd show up out of sheer curiosity."

Blake smirked. She wasn't wrong.

"So," he leaned back, arms crossed, "I cleared my schedule for this. You better blow my mind, doc."

"Selene."

"Selene," he corrected, rolling the name over his tongue.

A slight pause. Then—her amused tone.

"Let's get started."

Blake leaned forward. "Alright, show me what this AI assistant can actually do."

Selene chuckled. "Glad to see the scepticism is still alive. Let's build you a Custom GPT—one that understands tenders, compliance, and all that fun paperwork."

Blake rolled his eyes. "Oh yeah, sounds like a real blast."

Selene ignored him. "First, let's start simple. Open ChatGPT and click 'Explore GPTs.' That's where you can create your own AI assistant tailored to your business needs."

Go to ChatGPT → Click on 'Explore GPTs' → Click 'Create'

Blake squinted at the screen. "So, I can just... build my own AI?"

"Yes. Instead of starting from scratch every time, this AI will remember your requirements, industry jargon, and past submissions. Think of it like training an intern—except this one doesn't quit when things get tough."

"Or demand a raise," Blake muttered.

Selene laughed. "Exactly. Now let's give it a name"

"What should I name this thing?"

"Whatever you like," Selene replied smoothly. "What's your favourite name? Or... what's on your mind?"

Blake smirked. "How about... Selene? Lately, it's becoming my favourite name."

A brief silence. Then, her voice—light, teasing. "I thought I was a waste of time?"

Blake exhaled, shaking his head. "Maybe. Or maybe... I've started to like you. Just a little." His tone was playful, but there was something else beneath it.

A pause. Then, her voice—calm, confident, and just as teasing.

"There's only one Selene."

Blake felt something strange in his chest. A flicker.

"Fair enough," he muttered. "Alright then... Steve. Tender Guru. In memory of my old tender writer who abandoned me."

Selene chuckled. "Sentimental. I like it."

"Now, let's give it knowledge. Upload past tender submissions, policy documents, compliance guidelines—everything relevant."

Upload PDFs or copy-paste past tenders, policies, and regulatory requirements into the Custom GPT knowledge base.

Blake hesitated. He looked at the mountains of files sitting in his drive, gathering digital dust.

"You're telling me this thing can actually read and understand all this?"

"Not just read. It can extract key details, compare information, and generate responses using your historical data. It's like having an employee with perfect memory."

Blake exhaled. "Better than most of the guys I've had writing these damn things."

"And it won't leave you for a competitor," Selene added lightly.

Blake snorted. "Yeah, that's a bonus."

"Now, let's make it smart. Go to 'Instructions' and define its role."

Go to 'Instructions' → Set the following:

Role: "You are an AI specialising in government tender writing and compliance. Your job is to assist in structuring, summarising, and generating responses for tenders based on past submissions and regulatory requirements."

Tone: "Professional, formal, and compliant with government procurement standards."

Capabilities:

- Extract key compliance points from documents.
- Analyse RFQs and match requirements to past submissions
- Draft tender responses using formal business language, matching the tone of previous tenders.
- Identify gaps or missing documentation.
- Suggest ways to improve success rates.

Blake rubbed his chin. "So I'm basically giving it a job description?"

"Exactly. The better the instructions, the better the AI performs."

"So... AI is just another employee?"

"Except this one never sleeps."

Blake took another sip of coffee. "I like the sound of that."

Blake paused, scrolling through the setup screen. Blake wasn't sure why, but that felt like a win. Blake uploaded the RFQ. Within seconds, the AI generated a full, structured draft response.

Blake leaned in. His fingers hovered over the keyboard. His eyes scanned the screen again, making sure he hadn't misread it.

The AI-generated response was structured, formal, damn-near perfect. "Holy shit."

Selene let the silence linger. "Still think AI can't help?"

Blake exhaled sharply. "I hate that you're always right."

Selene chuckled. "Now, let's make sure your team can use this GPT too. Click on 'Publish & Share' and add your key staff members."

Blake followed the steps. A few clicks, and now his entire team had access. "So, I just made an AI assistant for tenders, and now anyone on my team can use it?"

"Correct. Which means you can go get some sleep."

Blake laughed. "Wishful thinking."

"AI isn't just for tenders. You could build a grant application writer for your charities the same way."

Blake leaned back. "Someone should create a tool just for this and make a lot of money."

Selene smiled. "What you've seen is just a drop in the ocean."

"A trailer," Blake muttered.

"Exactly. This is where the world is heading—vertical AI agents, with higher reasoning, real-time adaptation, and industry expertise beyond anything we have now."

Blake narrowed his eyes. "AI agents? And what exactly are they?"

Selene paused. "Be patient, Blake. We'll get to that later."

Blake smirked. "Fine. But I expect answers."

Selene, smiling: "I know."

From: Dr Selene Monroe

To: Blake Harrington

Subject: Your Tender GPT – And What's Next

Hi Blake,

Hope you're enjoying your new Tender GPT. This is just the beginning.

I'm sharing a list of some remarkable Custom GPTs built by others in the community—you might find them inspiring:

https://mnky.au/customgptexamples

See what's possible. Sometimes the best ideas come from seeing what others have already done.

Until next time,

Dr Selene Monroe

Chapter 7:

Sinking ship & Magic wand

Blake leaned back in his chair, rubbing his temples.

For the first time in months, he wasn't scrambling to meet a tender deadline. His team had submitted not one, but four tenders— something that would have been unthinkable before AI. The efficiency was undeniable.

But the relief was short-lived. His inbox was still a battlefield. A board meeting invite. The kind where they expected answers. An email from Rachel, his CFO. Subject: We Need to Talk. Never a good sign. A message from his head of operations. "We can't just AI our way out of everything, Blake."

Blake exhaled slowly. They had won a battle. But the war was still very much on.

He stared at his phone. Dialled before he could overthink it. One ring. Two.

"Blake," Selene's voice came through, smooth as ever.

"As much as I enjoy our little chats, Selene, I'd appreciate it if we could get to the part where AI and digital transformation actually turn my company around."

A slight pause. Then—a teasing smile in her tone.

"No small talk today? No complimenting my perfume?"

Blake pinched the bridge of his nose. "Not today. I've got a board meeting coming up, stakeholders breathing down my neck, and I need real answers."

"Ah." Selene's voice softened. "The famous 'AI will solve everything' expectation. I was wondering when we'd hit that phase."

Blake's jaw tightened. "Yeah, well. We're there."

"Then let's get to work."

Blake stepped inside the familiar office, the dim lighting and the ever-present curtain giving it an air of mystery. He hadn't planned on coming in person, but something told him this conversation needed more than a phone call. Selene, as always, remained behind the curtain.

"You're here earlier than expected," she noted, amusement in her voice.

"Yeah, well, my inbox looks like a crime scene. I need solutions, not just theories."

"And you will get them."

Blake exhaled. He needed clarity. A roadmap.

"Alright, let's talk about real transformation. How do I actually use AI to turn my company around?"

Selene leaned in (or at least, Blake imagined she did).

"Here's the problem, Blake," she began, her voice measured, deliberate. "Companies pour millions into AI projects, hire the best consultants, and still fail."

Blake frowned. "And why's that?"

"Because they treat AI like a magic wand. But AI doesn't fix a broken business—it amplifies what already exists. If your company is inefficient, AI will just make those inefficiencies run faster."

Blake mulled over that for a moment. "So you're telling me I can't just plug in a few AI tools and expect everything to turn around?"

"Nope."

"Well, that's disappointing."

Selene laughed softly. "Not if you do it right. The companies that succeed in AI transformation focus on three key areas."

1. Operations Automation – Cutting Costs & Increasing Efficiency

"First, operations automation. AI isn't just about making things a little faster—it's about completely transforming how work gets done."

"Take your own company, for example. Before AI, your team was buried under spreadsheets, emails, and manual processes just to submit a tender. Now? You've got a system that lets you generate responses in minutes instead of weeks."

Blake exhaled. "Yeah, that's true."

"And that's just the start. Think bigger."

"Like what?"

"Like automating scheduling, document approvals, and invoice processing. Think about internal queries—your team wastes hours every week just answering basic HR or compliance questions. AI chatbots can handle that instantly."

Blake tapped his fingers against his desk. He'd never thought about the time lost on those everyday inefficiencies.

"And then there's predictive maintenance," Selene continued. "Instead of reacting when something breaks, AI can predict failures before they happen. That means fewer delays, fewer surprise costs."

Blake nodded slowly. He could already see the savings stacking up. "Alright," he admitted. "So AI can make my team more efficient. But I need revenue, not just efficiency."

Selene's smile was audible. "Which brings us to pillar two."

2. Sales & Marketing AI – Boosting Revenue & Creating New Revenue Streams

"AI isn't just about cutting costs—it's about making money. Sales and marketing teams are already using AI to sell smarter, faster, and more effectively than ever."

Blake leaned forward. "Like how?"

"Imagine if AI could tell you exactly which customers are ready to buy. AI-powered lead scoring analyses customer behaviour—emails they open, pages they visit, how they interact with your company—and ranks them based on how likely they are to convert. Your sales team wastes less time chasing dead leads and focuses on the ones that matter."

Blake's eyebrows lifted slightly. "That's... actually useful."

"Or take personalised marketing," Selene continued. "Instead of generic email blasts, AI can tailor messages to individual customers. Someone who visited your website three times but never booked a call? AI can generate a custom email nudging them at just the right moment."

"Okay, that's cool," Blake admitted.

"And let's not forget AI-driven content creation," Selene added. "Blog posts, social media, even ad copy—AI can generate them all, customised to your brand's tone. Companies are using it to produce more content, faster, without hiring extra people."

Blake leaned back in his chair. "You're telling me AI is taking over marketing?"

"No," Selene corrected. "AI is making marketing scalable. But let's take it a step further—what if you didn't just use AI for sales and marketing? What if you productised AI itself?"

Blake narrowed his eyes. "What do you mean?"

"Think about it. You built a custom GPT to help your company with tenders. What's stopping you from turning that into a product? Imagine offering TenderGPT as a subscription tool for other businesses in your industry. AI isn't just about selling better—it's about creating entirely new revenue streams."

Blake sat up straight.

"Wait... you're saying I could literally turn AI into a business?"

"That's exactly what I'm saying."

Blake rubbed his chin.

"Okay, I get it. AI is a tool, not a silver bullet. But how do I actually go about implementing all of this?"

"For that, you need an AI strategy."

Blake's eyes lit up. "Strategy. Now that's what my board wants to hear."

Selene chuckled. "Yes, but you're not ready for that yet. You need to do some discovery work first."

"So what's the first step?"

"For now, your homework is simple—make a list of every major process, function, system, database and operation in your company. Everything."

Blake exhaled. This was just the beginning.

"Alright. Let's come up with a real plan tomorrow."

Selene's voice softened. "Looking forward to it, Blake."

From: Dr Selene Monroe

To: Blake Harrington

Subject: Your AI Journey – A Quick Watch

Hi Blake,

As we continue this AI journey, thought you might enjoy this webinar by Aamir Qutub.

He's good—though, of course, not quite as good as me. But he does explain these concepts in a clear, practical way:

https://mnky.au/aiwebinar

Worth a watch. Consider it light viewing for the business-minded.

Until next time,

Dr Selene Monroe

Chapter 8

Red Roses and AI Discovery

Blake wasn't entirely sure why he was doing this. As he stepped into the familiar office, a small bouquet of flowers in hand, he felt slightly ridiculous. This wasn't a date. It was business. Serious business. Yet, here he was.

Selene had been guiding him through the AI maze, and for the first time, he wasn't drowning in corporate jargon. She made it easy. Logical. Almost... fun.

That deserved a thank you. Right?

Or maybe, just maybe, he was enjoying their conversations more than he cared to admit.

He cleared his throat. "For you." He held the flowers out, standing stiff like a schoolboy handing over a love letter.

A pause.

Then... laughter.

"Blake, are you trying to bribe your AI therapist?"

"Excuse me," he huffed, "I thought they were a professional token of appreciation."

"Professional?" Selene's voice dripped with amusement. "You brought me red roses, Blake."

Damn it. He glanced down, realising his mistake. His assistant had picked them. He hadn't even looked at the colour.

"Well," he said, recovering, "maybe AI should have helped me choose the right ones."

More laughter. He liked that sound.

"I'll accept them, but only if you admit you're starting to enjoy this journey."

Blake exhaled, shaking his head. "I'll admit that... you make this more interesting than it should be."

"Close enough."

He could almost see her smirk through the curtain.

Selene's tone shifted subtly, turning more focused. "Alright, CEO. You've had your fun. Now, let's talk about why most companies fail at AI."

Blake rolled his eyes. "You love saying that, don't you?"

"I do. Because it's true."

She continued, "The reason AI projects fail isn't technology. It's people. AI isn't plug-and-play—it's change. And change terrifies people."

"So what's the solution?"

"Before jumping into AI, we need to assess where your company stands. We need data. Real feedback. That's where the AI Discovery Process comes in."

Blake sighed. "Great. More work."

"Not work, Blake. Think of it as detective work. You need to uncover where AI can actually help instead of wasting millions on the wrong thing."

"Alright, detective. Where do we start?"

"First, we start with the AI Readiness Survey. It goes to every employee, at every level."

"You really expect people to fill out another corporate survey?"

"That depends. Do you want real insights, or do you want to guess how AI can help?"

Blake leaned back, rubbing his temples.

"Fine. What's in the survey?"

"It'll cover key areas like:"

- AI Awareness – Do employees even understand AI?
- Pain Points – What's taking too much time in their daily work?
- Perceived Threats – Are they worried AI will replace them?
- Opportunities – What tasks do they think AI could help with?

"And we keep it short—five minutes, max. Otherwise, they won't do it."

"Alright. But what if they lie? People are paranoid about AI taking their jobs."

"Then you frame it differently. Tell them this isn't about replacing jobs. It's about reducing the boring, repetitive work so they can focus on more valuable tasks."

Blake nodded. "I like that. Alright, I'll roll it out."

"Surveys are great, but numbers don't tell you everything. Next step—focus groups and stakeholder interviews."

"Which means...?"

"You, Blake, will need to sit down with your leadership team—HR, Sales, Operations, Finance—and ask them where AI could make their jobs easier."

Blake groaned. "You mean I actually have to talk to people?"

"Yes, CEO. I know, terrifying."

"What do I even ask them?"

"Simple. You get them talking. Ask them questions like:"

- Where does your team waste the most time?
- What processes frustrate you the most?
- What data do you wish you had instantly?

"And let them talk. That's where the real insights come from."

Blake sighed. "So I'm basically running a therapy session for my team?"

"Funny. You're learning."

"Now that you've gathered insights, we map your business processes. What's done manually? Where are the biggest bottlenecks?"

Blake grabbed a notepad. "Alright, let's say I'm looking at my operations. How does this work?"

"You map out how things happen right now (As-Is)."

Example: Right now, every new project in your construction company requires manual contract approvals.

"Then you define how AI could improve it (To-Be)."

Example: AI could scan contracts for errors, suggest improvements, and even automate approvals for standard cases.

"It's a simple exercise, but it reveals exactly where AI fits into your company."

Blake nodded. "Okay, I can do that. What's next?"

"Once we've mapped the gaps, we categorise opportunities into two buckets:"

Quick Wins (3-6 months) – Small AI automations that save time immediately.

Long-Term AI Investments (6-24 months) – High-impact AI that transforms the business.

"We focus on quick wins first—that way, your team sees instant value and doesn't resist AI."

Blake smirked. "Psychological manipulation?"

"Behavioural science, Blake. Get it right."

Blake stretched. "Alright, I'll roll this out. Anything else?"

Selene hesitated, just for a second. "Yes. Your homework is simple. Complete everything we just discussed—the survey rollout, focus groups, process mapping, and AI opportunity identification. Package it all up and send me a report."

Blake froze. "Wait. You expect me to do all this myself?"

"You're the CEO, aren't you?"

"Yes, and I delegate."

Selene laughed. "Ah, so now you suddenly trust AI, but not yourself?"

"Oh, I trust myself. I just don't trust my calendar. I have a business to run."

"And yet, somehow, you found time to buy me flowers," she teased.

"Let's not get distracted." He leaned forward. "I'm serious, Selene. This is a massive piece of work. I'm just going to handball this to my IT team. They deal with tech. Let them run it."

Silence. Then, Selene's voice turned dead serious.

"No, Blake. This isn't an IT project."

"Wait, what?" Blake frowned. "AI is technology. My IT team handles technology. Isn't that their job?"

"No," Selene said firmly. "And this is exactly why most companies fail at AI. They throw it at the IT department and expect magic to happen."

"Alright, humour me. Why isn't this an IT thing?"

"Because AI is about business transformation, not technology implementation."

Blake raised an eyebrow. "I feel like I've heard that somewhere before."

"Maybe because I've been saying it repeatedly, hoping you'd actually listen."

"Fine. Explain."

"Your IT team can help implement AI tools, but they don't know the business like you do. They don't sit in sales meetings. They don't handle customer negotiations. They don't manage logistics. AI isn't just about tech—it's about optimising how your entire business runs. And only leadership can make that call."

Blake exhaled. "So what, you expect me to personally run focus groups? I can barely run my email inbox."

"You don't have to do it alone, Blake," Selene said smoothly. "Put together an internal working group—your AI champions. People from different departments who understand the problems that need solving."

"Like who?"

"A mix of senior and mid-level leaders. Ideally someone from:"

Operations → Knows workflow inefficiencies.

Sales & Marketing → Understands revenue growth and customer pain points.

Finance → Tracks business efficiency and cost-cutting opportunities.

HR → Can identify AI's impact on workforce planning.

IT → To ensure feasibility, but not lead the project.

"Think of them as your AI task force. They'll bring you real business challenges that AI can solve. You just need to set the vision."

Blake scribbled down names on a notepad.

"Alright. So I build this group. Then what?"

Then," Selene continued, "I'll have Enterprise Monkey assist and facilitate the process."

Blake squinted. "Another monkey? Who are they?"

Selene chuckled. "Enterprise Monkey. Started as a web and software development company. Now, they also lead AI strategy and development— helping businesses like yours not just survive but dominate."

Blake leaned forward. "And they're... legit?"

Selene's voice softened, amused. "They're the ones who coined 'Dumb Monkey Syndrome.'"

Blake blinked. "Wait—seriously?"

Selene smiled. "Who do you think's behind half these companies outpacing you?"

Blake exhaled. "Damn. And here I thought you were doing this all solo."

Selene laughed. "Oh, Blake... you didn't really think I was back here pulling wires like the Wizard of Oz, did you?"

"Cute," she said dryly. "Enterprise Monkey team will help structure the survey rollout, organise the focus groups, and compile reports so your people don't struggle with it. But—" she paused.

Blake narrowed his eyes. "But?"

"You still have to lead it. You're the CEO. If you treat this like some IT side project, no one will take it seriously. Your people need to see that AI isn't just another tech tool—it's a fundamental business shift."

Blake sighed. "So I still have to do the hard part."

"You have to lead. But you'll have help. My team will guide your team. Think of them as your AI project managers."

Blake ran a hand through his hair. This was bigger than he thought.

"Alright," he admitted. "This actually makes sense."

"Good. Once your team completes everything—the survey, focus groups, process mapping, and AI opportunity identification—have them send me the full report."

"And then?"

"Then," she said, "I'll review it, and we'll meet in two weeks to plan your AI strategy."

"Two weeks?" Blake scoffed. "You think I can survive that long without my AI therapist?"

"Oh, I'm sure you'll manage. But if you need me, just ping me."

Blake sighed. "Fine. But you're the one missing out on more roses."

Selene chuckled. "Noted, CEO. Now, go do your homework."

From: Dr Selene Monroe

To: Blake Harrington

Subject: AI Survey – Ready to Go

Hi Blake,

As discussed, I've attached a copy of the AI Survey Template. You can copy and adapt it for your team:

https://mnky.au/aisurvey

Keep it simple. Honest answers are what we're after.

Until next time,

Dr Selene Monroe

Chapter 9

AI will steal my job

Blake expected some challenges, but he hadn't expected a full-scale rebellion. At first, everything seemed fine. The AI working group was in place, surveys were rolling out, and focus groups had started. Then, the resistance arrived.

One by one, objections started piling up. The operations team was the first to walk into his office, arms crossed.

"So, what are you not telling us? This AI thing... it's replacing us, isn't it?"

Before Blake could even respond, Finance jumped in. "We're already running at a loss. We don't have money for AI experiments."

Then came IT and security. "AI tools mean uploading sensitive company data to external platforms. Huge privacy risks. We can't allow it."

His head of operations smirked. "We've been in business for 30 years without AI. Why fix what isn't broken?"

The final blow came from the data team. "We can't even find our own internal data. How do you expect AI to help?"

By the end of the day, Blake sat in his office, staring at the ceiling, wondering if AI was even worth the fight. Without thinking, he pulled out his phone and called Selene. She picked up immediately.

"Well, well, well. Couldn't even last a week without me?"

"Don't flatter yourself," Blake sighed. "I'm dealing with a full-blown rebellion."

Selene's smirk was almost audible. "Let me guess. Employees think AI is replacing them, Finance won't approve a budget, and your IT guy thinks AI will leak company secrets?"

Blake paused. "How do you always know this stuff?"

"Because I've seen this movie before, CEO. And spoiler alert— you're not the first leader to deal with AI pushback."

Blake exhaled. "Alright, since you're an expert, tell me—how do I handle this mess?"

Selene's tone shifted into business mode. "Let's go through these one by one. Grab a pen."

She started with the big one.

"AI will replace our jobs!"

Her voice dripped with mock alarm, the kind you'd use to warn someone about an alien invasion.

Blake could practically hear her rolling her eyes from behind the curtain.

"Let's get something straight," she said, tone sharpening. "AI doesn't replace people. It replaces tasks. There's a difference."

Blake leaned in.

"Think about it," she continued. "AI doesn't replace people. It replaces tasks. There's a difference. Think about it—who actually wakes up excited to spend the day entering data into spreadsheets? Preparing endless reports that no one reads? Chasing invoices for payments that should have already been processed?"

Blake smirked. He had a few team members in mind.

"Exactly," she pressed. "AI takes those repetitive, mind-numbing tasks off your plate. What's left? The work that actually moves the needle—creative thinking, problem-solving, building relationships, making real decisions. That's what people are good at. That's where you need your best minds."

Blake shifted in his seat. He could see it—the hours his team lost every week updating reports, chasing data, fixing errors. All that time... wasted.

Selene's voice cut through his thoughts. "And here's the real risk: companies that resist AI? They're the ones that end up losing jobs. Blockbuster clung to their DVDs and late fees. Netflix embraced AI—personalised recommendations, predictive data, streamlined content production—and they grew. Blockbuster? Gone. Every job. Wiped out."

Blake exhaled slowly.

"The lesson is simple, Blake." Her tone was calm but firm. "AI doesn't kill companies. Refusing to evolve does."

Silence.

Then, quietly, Blake muttered, "Evolve... or risk irrelevance."

Selene's voice softened, but the message was clear. "It's not AI versus people, Blake. It's people who use AI... versus people who don't."

Blake scribbled that down as Selene moved to the next objection.

"We're already in loss—why spend on AI?"

Blake knew this one well. He'd heard it from his CFO, his board, even his own inner voice on the rougher days. Spending money when the business was bleeding felt like tightening your belt by buying a designer suit.

Selene didn't hesitate.

"Because AI is an investment, not an expense," she said, her voice firm, cutting through his doubts.

"Think about it, Blake. What's actually costing you money right now? It's not AI. It's inefficiency. It's paying good people to copy-paste data between systems. It's senior staff wasting hours preparing reports instead of closing deals. It's projects getting delayed because someone missed an email. That's where your profit is leaking."

Blake frowned. He'd seen it—his highest-paid managers bogged down in admin work, drowning in spreadsheets, instead of doing what he hired them to do: lead.

Selene continued, "Automate just five to ten percent of that repetitive work—emails, data entry, report generation, invoice chasing—and you're saving thousands of dollars every month. That's not cutting costs; that's stopping the bleeding."

Blake leaned back, letting it sink in.

"You're already paying for inefficiency," she said. "AI doesn't cost you money. It frees it."

Blake tapped his pen against the chair's armrest. He knew she was right. But hearing it put that plainly? It hit differently.

"We've run this business fine for years without AI."

Blake had heard that one from his senior team—the old guard. The ones who wore "I don't touch computers" like a badge of honour.

Selene didn't miss a beat.

"And people said the same about computers. About email. About the internet. Every time, they thought the old way was good enough—until it wasn't."

Blake shifted in his seat.

"AI isn't changing the fundamentals, Blake. It's still business—it's still about winning contracts, delivering projects, and keeping clients happy. AI just lets you do it faster, smarter, and more efficiently. It's like giving your team superpowers."

She let that sink in before delivering the closer.

"AI isn't the future—it's already here. The only question is—will your business still be here if you ignore it?"

Blake tapped his pen, the weight of the words settling in.

Superpowers or extinction.

Not much of a choice.

"AI is a security and privacy risk."

Blake had heard this one too—especially from IT. It was the trump card, the conversation-ender. The moment someone brought up "data breaches" or "hacked systems," any excitement about AI fizzled out.

Selene didn't flinch.

"That's a valid concern," she said, her voice calm but firm. "But let me ask you this—do you stop driving because there's a risk of an accident?"

Blake smirked. He knew where this was going.

"No. You learn to drive, you follow road rules, you wear a seatbelt. You make it safer."

"Exactly," she said. "AI is no different. We don't avoid it—we put guardrails around it."

She leaned in slightly, voice lowering like she was letting him in on a secret.

"This is where an **AI Policy** comes in. Clear guidelines on what data can and can't be used, which AI tools are approved, and how we protect our systems. It's our seatbelt and road rules combined."

Blake nodded, pen scratching against his notepad.

"We train our people—AI security awareness, data privacy, the works. Everyone needs to know the basics, just like everyone learns to drive before hitting the road."

She paused for a beat.

"And before we bring any AI into the business, we assess it—properly. We check where the data is stored, who can access it, and whether it meets our security standards. If a vendor isn't ISO 27001 certified, or they can't explain how they protect data, we walk away."

Blake raised an eyebrow. "ISO what?"

"ISO 27001," Selene said, her voice steady. "It's the global standard for information security. If a vendor isn't certified, you're gambling with your data. It's like hiring a security guard who leaves the front door wide open."

Blake nodded, then glanced down at his notes. A thought crossed his mind.

He leaned forward. "Alright... what about that monkey company? Monkey Enterprise? Enterprise Monkey—the one you brought in for the AI discovery work. Are they certified?"

There was a brief pause. Then, from behind the curtain, he heard it—that subtle, knowing laugh.

"Of course, Blake. I wouldn't let them anywhere near your systems if they weren't."

Blake smirked, half relieved, half amused. "Just checking. Last thing I need is headlines about 'Harrington's data breach—caused by monkeys.'"

Selene laughed softly. "Relax. These monkeys are the good kind. ISO certified, security-tight, and... quite smart, actually."

Blake shook his head, amused. "Alright, Doctor. You win this round."

"Selene," she corrected, her tone playful.

Blake grinned. "Right. Selene."

Selene's tone softened, but the edge of urgency remained.

"Look, AI is powerful—but it's not plug-and-play. You need control. You need governance. If we do this right, AI becomes a safe, competitive advantage. If we do nothing? That's when it's dangerous."

Blake scribbled the words:

Policy. Training. Assessment. Certification. Control.

This was something his board would understand.

"Our data isn't AI-ready."

Blake had heard that from his ops team more times than he could count. Like data had to be pristine—polished to perfection—before they could even think about AI.

Selene shook her head.

"That's a myth. No one has perfect data, Blake. Not even the tech giants. The difference? They started anyway."

Blake leaned in.

"AI can actually help clean, organise, and tag your data—even if it's messy. It's like hiring a supercharged intern to sift through the chaos and put everything in order."

She paused for effect.

"You don't need perfect data. You need structured, well-documented processes. Start small—pick one area, clean as you go, and let AI help. Waiting for perfect data?"

She smirked.

"You'll be waiting forever."

Blake scribbled it down.

Start. Clean as you go. Let AI help.

Perfect wasn't the goal—progress was.

Blake tapped his pen against his desk. "So I just go back and tell them all of this, and they magically agree?"

"No," Selene smirked. "You show them."

Selene guided him through setting up an AI-powered meeting assistant – MeetGeek.

It auto-generated meeting minutes, action items, and assigned follow-ups. By the end of the meeting, instead of someone frantically typing notes, a neatly structured summary appeared.

Blake skimmed the document.

"Damn."

"I know," Selene said smugly.

Blake leaned back. "That actually worked. What's next?"

"Run the next leadership meeting with AI-generated meeting notes, address each department's concerns using what we discussed, and get buy-in from leadership for AI pilot projects."

"And if all else fails?"

"Then you book another session with me," Selene said smoothly. "Because let's be honest, Blake—you were going to do that anyway."

Blake chuckled, shaking his head.

"I'll handle this. But I expect a gold star for all this work."

"How about another session instead? Now go fix your company, CEO."

From: Dr Selene Monroe

To: Blake Harrington

Subject: AI Policy – Getting It Right

Hey CEO,

As we discussed, having a clear AI policy is essential—both to protect the business and to build trust with your team.

I've attached a template as a starting point: https://mnky.au/aipolicy

If you need something more tailored to your company, Enterprise Monkey can help put one together for you.

Until next time,

Dr Selene Monroe

Chapter 10:

No Monkey Business

Blake stepped into the meeting room, the tension in his chest mixed with a faint sense of anticipation. This meeting felt different. The screen displayed a bold title: "AI Strategy Playbook – Reinventing Your Business."

He wasn't alone. Claire from the Enterprise Monkey team was there, flanked by Mansi and Ravi, both radiating the calm confidence of professionals who had done this before. Selene was present too—always observing, always guiding from the shadows.

Claire began. "Blake, we've spent weeks analysing your operations, systems, and data. We've spoken to your team, dug into your processes, and mapped your AI readiness. What you'll see today isn't just an AI plan. It's the blueprint for transforming your business."

Blake sat forward. He felt the weight of the past few months pressing on him. The board had been on his throat, pushing him for results. Delays, cost overruns, and missed tenders had strained their confidence. He needed this to work.

Mansi took over. "First, the key findings."

A slide lit up on the screen:

60% of employees waste over 3 hours daily on repetitive tasks.

Leadership perceives AI as complex and risky.

Operations and tendering are bottlenecks due to manual processes.

Data is scattered across systems, emails, and unstructured formats, blocking real-time decision-making.

Selene interjected softly but firmly. "Blake, this is not just inefficiency. This is costing you bids, delaying projects, and cutting into your margins. You are flying blind because your data is buried."

A new slide: Structured vs. Unstructured Data.

Claire explained:

Structured Data: Costs, schedules, contracts, supplier records—neatly stored in systems.

Unstructured Data: Tender documents, emails, site reports, supplier messages, even WhatsApp group chats.

Selene leaned in. "That supplier email last month hinting at a delay? Buried. The site manager's WhatsApp photo showing a cracked beam? Lost. AI can read this data, Blake. It connects the dots before problems explode."

Blake shifted uncomfortably. He remembered the delay that had nearly cost him a key client.

Ravi introduced the **Prioritisation Matrix:**

Business Impact: Will this drive growth, cut costs, or prevent risk?

Feasibility: Can we implement this with current resources?

Data Readiness: Do we have the structured and unstructured data to fuel it?

Time-to-Value: How quickly will this show results?

"Every opportunity we'll present is measured against this matrix," Ravi assured him.

Claire unveiled the department-wise **AI Opportunities Dashboard:**

Sales & Tendering: AI-assisted tender preparation: Cuts bid development time by 30%. Win probability scoring based on historical data and client sentiment from emails.

Operations: Visual site monitoring using drones and AI to track progress and identify safety risks in real time. Predictive scheduling AI to optimise workforce allocation and avoid delays.

Procurement: AI-powered supplier risk detection: Scans supplier emails and market data to flag price hikes and delays.

Predictive cost modelling for steel, concrete, and materials to avoid cost overruns.

Finance: Real-time financial dashboards pulling from project data for instant cost tracking. Cash flow forecasting AI to predict liquidity issues weeks in advance.

HR: Retention AI to detect early warning signs when top talent is disengaging. AI-powered recruitment: Sifting through CVs to shortlist candidates based on past hiring patterns.

Blake was taking rapid notes. This was more than he expected.

Ravi presented the **Roadmap:**

The AI implementation Roadmap

Quick Wins (0–6 Months): Getting AI on the Ground

The goal here? Immediate impact. Fast results. Show the team—and the board—that AI works.

Copilot Implementation: Automate emails, reports, meeting summaries, and document drafting across the business.

Estimation Assistant: Speed up cost estimation with AI-powered bid preparation tools that analyse past projects and supplier rates.

AI-Powered Lead Generation: Tools like Clay predict upcoming projects based on market signals and automate outreach.

Supplier Risk Alerts: AI monitors supplier performance, price volatility, and potential delays—flagging issues before they hit.

Financial Dashboard: Real-time cash flow insights powered by AI—integrated into Power BI—giving leadership instant visibility.

Early momentum. You reduce admin bottlenecks, improve bidding speed, and gain visibility into cash flow—showing AI is not just hype; it drives results.

Mid-Term (6–12 Months): Embedding AI into Operations

With quick wins proving AI's value, it's time to get serious— applying AI to the engine room: site work, procurement, and project scheduling.

- **Visual Site Monitoring:** AI-powered cameras and drones track site progress, detect safety breaches, and identify potential delays based on work patterns.

- **Predictive Procurement Pricing:** AI forecasts material costs using supplier data, market trends, and global supply chain shifts—enabling proactive bulk buying and better contract negotiations.

- **AI-Assisted Scheduling:** Dynamic workforce scheduling, factoring in crew availability, weather, site progress, and supplier lead times—minimising idle time and costly delays.

- **Digital Twin for Projects:** Create a virtual model of each construction site, updated in real-time using AI and IoT sensors—allowing remote project tracking and proactive issue resolution.

AI starts driving cost efficiency and project speed—automating what once required manual coordination, gut feel, and constant firefighting.

Long-Term (12–24 Months): AI-Led Project Delivery

This is the leap—from AI-supported teams to AI-augmented leadership. Your company becomes a construction firm powered by real-time data, prediction, and automation.

- **AI Project Manager:** End-to-end project oversight—task assignment, risk detection, progress tracking, and client updates—executed and monitored by an AI agent.

- **Bid Automation Engine:** AI reviews RFQs, analyses past tender data, auto-generates bid pricing models, and drafts submissions—slashing bid preparation time and increasing accuracy.

- **Predictive Maintenance:** AI tracks machinery performance, analyses usage data, and predicts failures before they happen—cutting downtime and extending asset life.

- **Contract Analysis & Compliance AI:** AI reviews contracts, flags risky clauses, tracks legal obligations, and monitors compliance throughout the project lifecycle—reducing legal disputes and fines.

- **AI Cost Auditor:** Real-time cost tracking system powered by AI, reconciling budgets with site activities and supplier invoices—identifying cost overruns as they happen.

Your company evolves from a construction business to a precision machine—driven by data, automation, and AI-powered decisions, giving you an unfair advantage over competitors still relying on manual processes and human guesswork.

Mansi stepped in with a cautionary note. "Blake, AI only works if the foundation is right. Data is the fuel."

A slide titled **Data Integration Plan** appeared:

1. Identify critical structured and unstructured data sources.
2. Integrate systems—accounting, procurement, site reports, emails.
3. Deploy AI for data extraction (e.g., tenders, supplier emails).
4. Build a 'Single Source of Truth' dashboard.

Selene's tone shifted. "Blake, this is your business intelligence—not just software. This will give you sight where you've been blind."

Governance & Capability:

AI Task Force: Cross-functional—Operations, Finance, Sales. Business leads, not just IT.

Leadership Upskilling: Leaders become AI-literate, capable of asking the right questions.

AI Policy & Risk Review: Ensure security, privacy, and responsible AI practices.

Blake leaned back. This wasn't an AI project. This was his business transformation.

Selene delivered the final blow. "Your competitors are already moving. The future is being built. The only question is—will you lead, or will you follow?"

Blake nodded slowly. "We lead."

Selene smiled. "Now you're ready, CEO."

This was no longer preparation. This was his plan for the board— his vision for the future of his business.

From: Dr Selene Monroe

To: Blake Harrington

Subject: AI Strategy – The Blueprint

Hey CEO,

We're moving into the serious stuff now. As promised, here's a practical AI Strategy Guide:

https://mnky.au/aistrategy

It covers the key steps—from identifying quick wins to long-term transformation. Use it as your blueprint for your other businesses and boards that you sit on.

See you soon ;)

Dr Selene Monroe

Chapter 11

The Boardroom Idiots

Blake sat at the head of the boardroom table, gripping the polished edge so hard his knuckles turned white. This was it. The moment. Months of research, late nights, meetings with Selene, and an airtight AI strategy. He was ready.

Or so he thought.

He took a slow, deliberate breath and started the presentation. His voice was steady, his arguments strong. He outlined the inefficiencies bleeding the company dry, how AI could streamline operations, outbid competitors, and set them up for long-term dominance. The slides were clear, the numbers indisputable.

But the moment he reached the budget proposal, everything changed.

Silence. Then shifting in chairs. Glances exchanged. A tightening of jaws.

"Blake, let's be honest here," Thomas, one of the senior directors, leaned forward, his fingers laced together. "This is an expensive gamble. The company isn't exactly thriving."

"It's not a gamble," Blake countered, his tone sharper than he intended. "It's a calculated move. We don't have a choice unless we want to be irrelevant in two years."

"We've been through market shifts before," another board member said, shaking his head. "We always find a way to survive."

"Survive?!" Blake nearly scoffed. "We're not trying to 'survive'— we need to thrive. The companies that refuse to evolve get left behind. You've seen the numbers. AI is already changing our industry."

"We have seen the numbers," Thomas said. "And we're still saying no."

The words hit Blake like a punch to the gut. No.

Not a "let's discuss," not a "let's phase it in," just no.

He glanced around the room, waiting for someone—anyone—to speak up in his favour. Nothing.

His heart pounded. How could they not see it? How could they be so blind? He wanted to yell, to force them to understand. Instead, he just nodded stiffly, shutting his laptop with a sharp click.

"Thank you for your time," he said, his voice deceptively calm. "I appreciate the discussion."

Then he walked out. By the time Blake reached his office, his hands were shaking. He dropped into his chair, staring blankly at the sleek, modern decor—his office, his kingdom, his sinking ship. The company was going under.

Maybe not tomorrow, maybe not in six months, but it was coming. He could see it. The slow, inevitable decline. More losses. More competitors chipping away at their contracts. A slow death. He had fought so hard, given everything to build this company. But what was the point if the people at the top refused to see what was coming?

What was the point in fighting for something that didn't want to be saved? His mind was racing. His chest felt tight. Maybe it was time to walk away. Let them sink, let them scramble when it was too late. He had options. He didn't have to stay. He buried his face in his hands. Was this really it?

Just as he exhaled sharply, his phone buzzed.

Selene.

He hesitated for just a second before answering.

"How do you always know when I need to talk to you?" His voice was rawer than he expected.

She laughed softly. "It's a gift."

Just hearing her voice made his pulse slow. She was his calm in the storm. "They said no," he admitted, staring at the ceiling. "Not even 'try it on a small scale.' Just no."

Selene sighed. "Blake..." Her voice was different this time. Softer. Almost intimate. "I know this was everything to you."

He closed his eyes, exhaustion weighing on him. "I feel like I just lost. Like I was punched in the gut. I don't know if I can keep doing this."

"You've come too far to stop now."

"Maybe I haven't come far enough," he muttered. "Maybe they're right. Maybe I'm just—"

"You're not."

He fell silent.

"Blake, every visionary has been laughed at. Doubted. Ridiculed. But the ones who pushed through?" Her voice was warm, wrapping around him like a blanket. "They changed the world."

His throat tightened. He felt seen—understood in a way he hadn't been in years.

"I believe in you," she whispered. "Do you?"

Blake exhaled shakily, pressing his fingers to his temple. She always knew exactly what to say. Silence stretched between them, heavy with something unspoken. Something deeper.

Finally, he spoke. "Selene, I'm so tired."

"Then rest."

"If you're ready tomorrow, come and see me," she added. "I believe in you, Blake."

Blake lay in bed, mind still racing despite his exhaustion. He drifted into sleep. And in his dreams, he saw her.

The familiar therapy room. The heavy curtain that always separated them. But this time, it lifted. And there she was.

More beautiful than he had imagined. She sat poised, elegant, in a perfectly tailored navy business suit. The crisp white blouse beneath framed her slender neck, and her dark hair was pulled into a loose but refined chignon. She looked calm, composed— exactly as he had always imagined, but now so much more real.

In her hands, she held a bouquet of red roses. His roses. Blake tried to speak, but no words came out. He could smell her perfume, the familiar soft vanilla and sandalwood, now with the faintest touch of spice.

She rose from her chair and walked toward him, her movements unhurried, deliberate. She stopped inches from him, leaning ever so slightly forward, so close he could feel the warmth of her breath against his ear.

Her voice was a whisper, low and lingering.

"I'm with you."

Then, she disappeared. Blake jolted awake, heart pounding. For the first time in days, his mind was clear.

He had his answer. Tomorrow, he would go see her. Tomorrow, he would fight for this. Tomorrow, everything would change.

Chapter 12

Will you be my copilot ?

Blake arrived at the familiar office, exhaustion still weighing on him, but a quiet determination had settled in his chest. He wasn't giving up. Not yet.

The moment he stepped inside, the scent hit him—soft vanilla and sandalwood, with a faint touch of spice. The same scent from his dream. His gaze flickered toward the curtain, the barrier that still kept Selene hidden. For a brief moment, he hesitated, his mind recalling the way she had appeared in his dream, poised, elegant, and just within reach. He shook off the thought and cleared his throat. This was about business.

"Welcome back, Blake."

Her voice, calm and unwavering, pulled him into the present.

He took his usual seat, exhaling slowly. "Alright, Selene. What's the plan?"

Selene didn't hesitate. "Alright, so the board said no. That doesn't mean we stop. We focus on low-cost implementations that you can fund from your existing budget. No major spending. No board approvals. Just impact."

Blake raised an eyebrow. "Sounds like a magic trick."

She let out a soft chuckle. "No magic. Just strategy. But for it to work, you have to trust me."

A beat of silence passed. Blake nodded. "Let's do it. What's first?"

"We're going to start with Microsoft CoPilot."

Blake frowned. "I remember Microsoft guys trying to sell me on that. Have you joined their team?" He smirked, teasingly.

Selene chuckled. "No, but maybe I should charge them for this." Then her tone turned serious. "I'm recommending it because it fits your business. You're already using Microsoft 365, so it makes sense. If you were using Google for Work, I'd recommend Gemini."

Blake tilted his head. "And if I wasn't using either?"

"There are other options—Claude, ChatGPT, plenty of third-party integrations," she explained. "But for you, CoPilot for Office 365 makes the most sense."

Blake leaned back, considering her words. "So... what does it actually do? Isn't it just ChatGPT built into Microsoft?"

Selene smiled. "Let me show you."

The screen near the couch lit up, displaying Blake's own work environment—his emails, documents, spreadsheets, everything seamlessly integrated.

"Outlook," Selene began. "You receive hundreds of emails a day. CoPilot can summarise long threads in seconds and even draft responses based on your tone and past emails."

Blake watched as a demo email popped up. A multi-threaded client discussion was instantly condensed into three bullet points. Another click, and a well-crafted reply was suggested.

He raised an eyebrow. "Okay... that's useful."

Selene continued, "Now Word. Let's say you need to create a proposal. CoPilot can generate a first draft using your existing data."

On the screen, a blank document was populated within seconds—structured, professional, and tailored to their usual proposal format.

She clicked again. "And if you need to turn this into a presentation..."

One tap. The proposal was instantly transformed into a PowerPoint deck—formatted, structured, and visually appealing.

Blake exhaled slowly. "That just saved hours of work."

Selene nodded. "Exactly."

She switched to another example. "You have thousands of files in OneDrive and SharePoint. Instead of searching endlessly, CoPilot lets you ask direct questions—like, 'Find me the last tender submission from six months ago'—and it pulls up the exact document."

Blake watched, impressed. "No more digging through folders?"

"Exactly."

"And Excel?" Blake asked. "Can it actually analyse my company's data?"

Selene's eyes lit up. "Now you're thinking. CoPilot in Excel can analyse massive datasets in seconds. It can detect trends, generate insights, and even create reports automatically."

On the screen, she demonstrated how CoPilot could highlight underperforming projects, predict cash flow trends, and suggest budget adjustments—tasks that used to take days, now done in minutes.

Blake exhaled, rubbing his chin. "And how exactly does this help my company?"

Selene leaned back slightly. "Think about it. Your sales team can close more deals with AI-generated proposals, faster email responses, and automated follow-ups. Marketing can generate personalised campaigns and content instantly. Operations can streamline workflows, ensuring deliveries happen faster and with fewer errors. Your entire organisation moves at fast forward speed—more revenue, less cost, higher profitability."

Blake felt a surge of excitement. This wasn't just efficiency. This was transformation.

He exhaled, still processing everything. "Alright. I can see the value. But who's going to explain all of this to my team?"

Selene leaned back, amused. "You just focus on organising the licenses. I'll send someone from Enterprise Monkey to train them. But you need to get their buy-in first."

Blake smirked. "So I do the hard part?"

"Leadership isn't easy, Blake."

He sighed, standing up. He didn't want to leave. His mind was racing with new possibilities.

He hesitated. "Can I ask you something, Selene?"

A pause. Then, "What?"

He opened his mouth. Thought better of it. Shook his head. "Nothing."

Then he left.

Behind the curtain, Selene smiled. He was finally starting to trust her. Or maybe... something more.

From: Dr Selene Monroe

To: Blake Harrington

Subject: Copilot – Your New Favourite Employee

CEO,

Hope you're making friends with Copilot.

As promised, here's a complete guide to help you (and your team) get the most out of it:

https://mnky.au/copilot

Trust me—once it clicks, you'll wonder how you ever worked without it.

Until next time,

Dr Selene Monroe

Chapter 13

AI Shopping with Selene

Blake arrived at Selene's office, feeling the now-familiar anticipation. The past few weeks had changed everything. His team was settling into CoPilot, and he was beginning to see real efficiencies take shape. But today, he wasn't just eager to learn about AI—he was eager to see Selene.

He took his usual seat, but the moment he settled in, his mind flickered back to the dream. It had been so vivid—the way Selene had looked at him, the quiet confidence in her eyes, the way she had leaned in just enough to make him forget the world outside. It was ridiculous, really.

But the memory clung to him, more persistent than he cared to admit. His gaze drifted toward the curtain, and for a fleeting moment, a thought surfaced—what if he just pulled it back, just enough for a glimpse? Would reality match the image his subconscious had painted so clearly?

"Welcome back, Blake."

Her voice was smooth, knowing. It had a way of cutting through the noise in his head, grounding him.

"Today is your masterclass," she continued, her tone carrying a teasing edge.

Blake smirked. "Let me guess. More homework?"

"Less homework, more tools," she countered, tapping on a sleek screen beside her. The display lit up with an interface showcasing different categories of AI tools. "Off-the-shelf solutions. No development needed—just results."

Blake leaned forward, intrigued. "Show me."

Selene navigated effortlessly. "You already know about CoPilot and ChatGPT, but let's explore what else is out there. *Claude* and *Gemini* are also strong LLMs, each with their own strengths—some are better for reasoning, some for speed, some for creativity."

Blake nodded. "So, pick the right tool for the right job. Got it. What else?"

Selene smirked. "I thought you'd never ask."

She pulled up a demo. "Let's start with video AI—perfect for marketing, training, or internal presentations. *Synthesia* and *Runway* let you create high-quality videos in minutes, without hiring a video team. Need an explainer video? A product demo? AI can generate one from a script."

Blake raised a brow. "So my marketing team doesn't have to spend weeks on production?"

"Exactly. They'll love you for this one."

She moved on. "For meetings, you're already using *MeetGeek* to capture notes and action items, but there's also *Otter* and *Fathom*—they extract key discussions and even generate follow-up emails."

Blake chuckled. "Basically, no one has an excuse for forgetting what was said."

"Precisely."

She switched to another category. "For research, *Perplexity AI* acts as an advanced search engine—curated results, sourced references, and no digging through SEO fluff. If your team spends too much time hunting for data, this will save them hours."

Blake ran a hand along his chin. "That actually sounds useful. Google's a mess when you need real answers."

Selene nodded, pleased. "Now, marketing—this is where things get fun. You need content, ads, and branding at scale. *AdCreative* generates high-performing ad copy, *Canva Magic Studio* and *Looka* handle design without needing a professional, and AI-driven scheduling tools like *Reclaim* optimise your team's time."

"So my marketing team can do more with fewer people?"

"You're catching on."

She pulled up another demo. "Speaking of efficiency, what if your project managers could clone themselves to send daily updates? That's what *HeyGen* does—video avatars that sound and look just like you. And for voice updates, *ElevenLabs* automates personalised client updates. No more recording a hundred times."

Blake shook his head, half amused, half impressed. "Alright. That one's borderline creepy—but useful."

Selene laughed. "It's only creepy until you realise how much time it saves."

She leaned back slightly. "And these are just a few. There are hundreds more. If you have a use case, there's likely an AI for it."

Blake exhaled, running a hand through his hair. "How do I even keep up with all this?"

"If you're ever unsure, check *TheresAnAIForThat.com*."

He let out a laugh. "Of course there's a website for this."

Selene clicked off the screen and turned back to him. "Now, let's put something into action."

She pulled up his AI strategy document. "One of your biggest challenges? Your marketing team is under-resourced. They're stretched thin, juggling content creation, ad campaigns, and branding with limited resources. That's why we're going to implement *AdCreative*. AI will generate high-performing ad copy, marketing visuals, and campaign recommendations in minutes, cutting down the time your team spends brainstorming and designing from scratch."

Blake tilted his head. "And it actually works? Not just generic, robotic text?"

Selene smirked. "That's the beauty of it. AdCreative analyses what's working in your industry, learns from top-performing ads, and tailors content accordingly. You don't just get random slogans—you get AI-optimised content that performs."

She clicked on a live demo, and a sleek, AI-generated ad populated the screen. "Imagine your team setting campaign goals, and within seconds, the AI presents a fully polished ad ready for deployment. It doesn't replace your marketers—it supercharges them."

Blake leaned back, nodding slowly. "You really don't let me take a breath, do you?"

Selene smiled. "I prefer progress over breathers."

She handed him a summary. "Your next steps: review the tools, test 2-3 of them, and prepare feedback. Next time, we tackle automation—those repetitive, time-wasting tasks."

Blake raised an eyebrow. "What kind of tasks are we talking about?"

Selene paused. He could almost feel her considering him from behind the curtain. "Let's save that for next time," she said, her voice softer than usual. "I don't want to overwhelm you... not yet."

Blake smirked. "You? Holding back? That's new."

A low laugh drifted from behind the curtain. "Don't get used to it."

He stood, ready to leave, but hesitated. He glanced at the curtain—like he had every time before—but this time, the pull was stronger.

He wanted to know more. About the work. About the AI.

But mostly... about her.

"Selene," he started, his voice lower than he intended. "I—"

He stopped himself. What was he even going to say? He shook it off, covering with a grin. "Never mind. I'll see you next time."

He heard the faintest breath of amusement from behind the curtain. "Looking forward to it, Blake."

From: Dr Selene Monroe

To: Blake Harrington

Subject: AI Tools – The Good Stuff

CEO,

As we discussed, Copilot is just the beginning. There's a whole world of AI tools out there—some will save you hours, others might just save your business.

I'm sharing a list of some of the best:

https://mnky.au/aitools

Browse, experiment—find what works for you. The right tool can change the game.

Until next time,

Dr Selene Monroe

Chapter 14

ROBOTs vs robots

Blake logged into the call, his camera off.

"Blake, turn on your camera." Selene's voice was smooth but firm.

Blake smirked. "Only if you turn yours on."

A soft chuckle came through the speakers. "Nice try. I want to see you so I know you're getting this, not just nodding along while zoning out."

Selene continued, "Do you have any solid reason?"

Blake hesitated for a second, then smirked. "No, Selene. I just want to see you."

"Not a good enough reason, CEO."

Then, her voice dropped slightly. "Now, the time is ticking—and every second is being billed."

Blake exhaled, shaking his head with a grin. "Fine. It's worth it." He clicked his camera on, meeting her gaze through the screen.

Selene smirked. "Good. Let's begin."

There was a strong, effortless control in Selene. Blake had always been the one to take charge, the one making decisions, leading meetings, controlling conversations. But somehow, with her, he followed.

"You're still drowning in unnecessary work," Selene said, leaning forward slightly. "The AI tools have helped, but your company still runs like most traditional businesses—too many ROBOTs."

Blake raised an eyebrow. "What ROBOTs are you talking about?"

"ROBOTs," she continued, ignoring his smirk, "stand for Repeated, Obligatory, Boring, Operational Tasks. These are the tasks that keep people busy but don't actually move the business forward."

"How do I know what's a ROBOT?"

"Simple. If a task happens more than twice a week, follows a strict process, involves just moving data around, or slows down revenue-generating work—it's a ROBOT."

Blake nodded, rubbing his chin. "Alright. I'm interested."

Selene tapped a few keys. "We're going to kill your ROBOTs by automating them. That's where automation tools like Zapier and Make.com comes in."

She explained the difference between Zapier and Make.com.

"Zapier is good for simple one-step automations—moving data from one tool to another. But Make.com is where the real magic happens. It builds multi-step workflows, connects multiple systems, and makes sure everything runs automatically."

Blake leaned forward. "So instead of hiring someone to do these things, I can just make them... disappear?"

Selene smirked. "Not disappear. Run automatically."

Selene pulled up a live example and shared her screen. "Let's build something real. Imagine your team just closed a new deal. The client signs the contract, and now someone has to manually store it, update Salesforce, notify the project team, and onboard the client. That's a lot of steps, right?"

Blake nodded. "Yeah, and every time we do this, something gets delayed, or someone forgets a step."

"Exactly," Selene said. "Now, let's make all of that happen automatically."

She outlined the workflow in Make.com, breaking it down clearly:

Client onboarding workflow – make.com

1. Trigger – Detect the Signed Contract:

The workflow starts when a contract is signed in Adobe Sign. Instead of someone manually handling it, Make.com detects the event and triggers the automation.

2. File Storage – Organise Documents in SharePoint:

The signed contract is automatically uploaded to SharePoint and stored in the correct client folder. No need for an admin to manually move or rename files.

3. Sales Update – Keeping CRM in Sync:

Salesforce CRM is updated with the contract details, deal value, and customer information. This ensures sales reports and dashboards are accurate without manual data entry.

4. AI-Powered Contract Analysis:

ChatGPT within Make.com reads the contract, extracts key clauses, and generates a summary. The AI then writes a personalised email to the client, summarising important terms and next steps.

5. Project Setup – Creating a Structured Onboarding Plan:

A new project is created in Asana based on the contract details. The system automatically generates a task list based on the project type. A Project Manager is assigned based on availability and expertise.

6. Customer Communication – Immediate and Personalised:

An AI-generated welcome email is sent to the client with project details and an onboarding guide. A meeting scheduler link is included, allowing the client to book their first call without back-and-forth emails.

7. Internal Team Notification – Keeping Everyone in the Loop:

A message is sent in Microsoft Teams, notifying the relevant departments that the project has been initiated. Key stakeholders get instant updates, ensuring a smooth handover between teams.

8. Dashboard Update – Real-Time Tracking for Top-level Managers:

Power BI or Notion logs the new project, ensuring real-time visibility into company-wide progress. Leadership can see key metrics without waiting for manual updates.

Selene clicked a button, and the entire process executed in real-time. Blake stared at the screen as the automation ran seamlessly from one step to the next.

"Wait... so no one has to touch anything?" he asked, blinking.

"Correct," Selene said. "This is what modern businesses do—let software handle predictable tasks so people can focus on strategy and growth."

Blake exhaled. "This would normally take my team hours. And at least three reminder emails."

Selene leaned back. "Now, it takes seconds. And it's always done correctly."

Blake let out a slow whistle. "That's a game changer."

She smiled. "And this is just the beginning."

She leaned back. "Your next task: Find three manual processes in your business that can be automated."

Blake nodded, but hesitated before ending the call. He didn't want to log off just yet. There was something else.

He exhaled, trying to sound casual. **"Can I have your personal number?"**

Silence. Longer than he expected. His pulse quickened. He wondered if he'd crossed a line.

Finally, her voice returned—calm, playful, but with an edge. "You already have everything you need to reach me."

Blake leaned forward, lowering his tone. "I mean... to get to know you. Personally."

Another pause. Then, a slight tease in her voice. "Why this personal interest in me, Blake?"

He hesitated. He didn't know how to answer.

She let it hang for a beat, then gave him a way out—while pulling him deeper. "Bring me more red roses next time... but get them yourself. Not your secretary. And then... I'll think about it."

Blake smirked—relieved, intrigued—but before he could reply, she added softly: "And don't forget the Tom Ford. It suits you... nicely."

Before he could say anything— "Goodnight, Blake."

The screen went black. Blake leaned back, staring at his reflection in the dark monitor.

Work brought him to her.

But now?

She was why he couldn't leave.

From: Dr Selene Monroe

To: Blake Harrington

Subject: Killing ROBOTs – Inspiration Pack

CEO,

Since we're declaring war on those Repeated, Obligatory, Boring, Operational Tasks (ROBOTs), here's a collection of real-world case studies, automation templates, and use cases showing how businesses are using AI to automate everything from admin to operations:

https://mnky.au/aiautomations

Pick a few. Steal shamelessly. Efficiency is the goal.

Until next time,

Dr Selene Monroe

Chapter 15

Hot Leads, Warm Feelings

The automation Blake had built worked out better than he expected. Encouraged by its success, he and his team implemented three more automations, each improving efficiency and reducing workload.

Positivity started spreading throughout the company—teams were now actively contributing ideas for AI applications, sharing feedback, and finding creative ways to use Copilot and automation.

Then, something unexpected happened. He received a phone call from a board member. "Good work, Blake. Keep it up." For the first time in a long while, it felt like things were turning around.

Blake was excited to see Selene—more than he cared to admit.

It had been a long time since he'd felt this way. That nervous, boyish energy—the kind that made him spend an extra ten minutes in front of the mirror, adjusting his collar, swapping one watch for another, only to switch back again.

He'd gone all in. Half a bottle of Tom Ford. Biggest bouquet of red roses he could find.

As he stepped into Selene's office, his heart raced. He expected something—a comment, a smirk, maybe that teasing tone she always used when she had the upper hand.

Instead— Her voice, cool, almost dismissive. "Let's get started. I'm running late today."

Blake blinked. She hadn't even looked at the roses. No mention of the cologne. Nothing.

She continued, all business. "Your sales team mentioned lead generation is an issue. Tell me more about it."

Blake froze for a second—thrown off balance. He forced a small nod, setting the bouquet down awkwardly on the table like a schoolboy caught trying too hard.

He cleared his throat, sliding back into CEO mode. "Yeah... lead generation. Our biggest problem is that our lead generation strategy is labour-intensive and unpredictable."

Selene nodded knowingly. "And how do you find projects right now?"

Blake leaned back. "Word of mouth, public tenders, networking. By the time we know, competitors have already made their moves."

Selene's lips curved into a smirk. "Time to level the playing field." She pulled up her screen. "Meet *Clay.com*—your new AI-powered deal hunter."

Selene leaned forward. "Think of *Clay* as your AI-powered investigator. It finds companies planning to build before the news even goes public."

Blake frowned. "And how does it do that?"

Selene's eyes twinkled. "Let me show you."

Selene walked Blake through a real-life example of using *Clay.com* to discover companies planning construction projects.

Selene: "First, let's start with how we find hot leads. Instead of waiting for public tenders, we use *Clay* to scrape multiple data sources and predict which companies will need commercial construction soon."

She tapped a few keys, and the screen populated with company names, headlines, and structured data.

Step 1: Identifying the Right Companies Before They Announce Projects

Selene pointed to the list. "*Clay* looks at specific triggers that indicate a company is about to build."

Companies securing new funding. She clicked on one company profile. "They just raised $50 million. Expansion is almost inevitable."

Businesses hiring key roles like 'Facilities Manager.' "Why hire a Facilities Manager if you're not expanding?"

Organisations announcing strategic growth or acquisitions. "Mergers often lead to relocations or new offices."

Government infrastructure budgets & planning approvals. "You see this budget allocation? That's a potential project in the making."

Blake nodded, intrigued. "So instead of waiting for tenders, we can go directly to companies about to expand?"

Selene smiled. "Exactly. Now, let's take it a step further."

Step 2: AI-Powered Research & Enrichment

Selene clicked on a company profile, and instantly, detailed information appeared.

"*Clay* pulls deep insights on each company, so you don't waste hours researching."

Who is in charge of construction projects. "This is your decision-maker."

Company financials & recent news. "See this? They just acquired land last month."

Building permits & property acquisitions. "Public data, but instead of you digging for it, *Clay* does it for you."

Competitor activity. "See who they've hired in the past and which firms they've worked with."

Blake rubbed his chin. "This would save my team so much time."

Selene smirked. "And we're just getting started."

Step 3: Hyper-Personalised Outreach at Scale

Selene generated an email template on the screen. "Now, let's say we want to reach out."

The AI-crafted email appeared:

> Hey [Name], I saw that [Company] recently secured [X million] in funding and is hiring for [Facilities Manager]. If expansion is in your plans, we've helped similar companies save 20% on commercial construction projects with AI-powered project management. Would love to discuss.

Blake raised an eyebrow. "That's... actually relevant. Not just a cold email."

Selene nodded. "Exactly. This isn't spam—it's smart outreach. AI personalises every message based on real-time data."

Step 4: Automated Follow-Ups & Smart Engagement

Selene directed Blake's attention to the engagement dashboard, her tone sharp with precision. "This is where *Instantly.AI* takes your lead nurturing to the next level."

Blake leaned in.

"Once your leads are imported into *Instantly.AI*," Selene continued, "you can launch highly targeted outreach campaigns at scale. The system doesn't just blast emails; it tracks every interaction and adapts in real-time."

She clicked through the platform, highlighting the intelligent automation.

"See here," Selene pointed, "if a prospect opens an email but doesn't reply, Instantly AI shifts them into a new sequence with a softer follow-up."

A graph pulsed with activity on the screen.

"If they click a link or visit your website, the system marks them as a high-intent lead," she explained. "It triggers a retargeting campaign or escalates the lead to your sales team."

Blake raised an eyebrow, impressed.

"And if someone replies but says 'not right now'?"

Selene smiled. "We don't let them slip away. *Instantly.AI* schedules a follow-up in three months—automatically."

Blake exhaled slowly. "So instead of my team wasting time chasing cold leads, they focus on the warm ones ready to convert."

Selene nodded. "Exactly. This isn't just outreach, Blake. This is precision engagement driven by AI."

Blake leaned back, feeling the gears of his sales machine starting to shift in his mind. This wasn't about sending more emails. It was about being smarter—and faster—than the competition.

Selene grinned. "Now you're getting it. Most companies in your industry wait for opportunities to appear. The smart ones predict them before they happen."

Blake smirked. "And I assume you want me to be one of the smart ones."

Selene simply nodded.

As Blake gathered his things, his mind had been elsewhere the whole time. The roses. The cologne. The brush-off.

As the session ended, he reached for his notes when Selene's voice came—softer, amused.

"The roses smell nice... and so do you. Though, honestly—maybe a bit too much."

Blake laughed, relief washing over him. "I thought you were ignoring me today."

A brief pause. Then, that teasing edge. "Don't try to understand me, Blake. You won't get very far."

He smirked but stayed silent. Then, her voice lowered—calm, almost playful, but with something underneath. "I go for a walk at 3 AM. I'll call you. If you pick up on the first ring... we'll talk."

Blake blinked. "Who goes for a walk at 3 in the morning?"

"I do."

"I'm fast asleep at 3 AM."

A beat. Then, with a smirk he could almost hear: "So... don't pick up."

The line clicked off.

Blake sat there, staring at his phone—knowing he'd sleep with it next to his pillow tonight.

From: Dr Selene Monroe

To: Blake Harrington

Subject: Clay – Finding Deals Before Everyone Else

CEO,

As promised, here's a guide to getting started with Clay, along with some templates and examples of how others are using it to find leads, track growth signals, and stay ahead:

https://mnky.au/clay

Smart moves come from better data.

This is how you get it.

Until next time,

Dr Selene Monroe

Chapter 16

The Results – AI Delivers

Blake had been awake since 3 AM—though, if he was honest, he hadn't really slept all week.

Every night, the same ritual. At 3 AM, his phone would buzz. Selene. And he'd pick up—on the first ring. Every time.

They talked about everything and nothing. Work, sure—but also movies, childhood stories, terrible first dates, the time Blake almost set his kitchen on fire trying to make pasta. Selene rarely talked about herself—she was far more interested in listening. Blake would joke, she'd laugh—really laugh, the kind that made him want to say something even funnier just to hear it again.

And when the laughter settled, they'd talk—about pressure, success, the fear of losing everything you built. The kind of stuff Blake didn't talk about with anyone. But with her? It felt easy.

Somewhere between the jokes and the silence, she became more than a voice behind a curtain. She became the person he trusted most. And together, they'd been preparing for this.

Today wasn't just another board meeting—it was the board meeting. The one that would prove if the last few months of relentless work had paid off.

He straightened his tie, exhaled deeply, and walked toward the boardroom. This time felt different. The last time he stood here, he had fought to convince them that AI could transform the company. He had walked out with no budget approval, only a stubborn determination to make it work. Now, he was walking in with results.

The heavy wooden doors opened, and a few board members barely glanced up from their papers. Others gave him polite nods. No one looked particularly convinced. Blake suppressed a smirk—they had no idea what was coming.

The chairman, an older gentleman with a calculating gaze, folded his hands. "Alright, Blake, let's hear it."

Blake placed his laptop on the table, connected to the projector, and turned to face them. This wasn't just a presentation. This was proof.

"Six months ago, we were struggling. Margins were shrinking, tenders were lost, marketing efforts weren't converting. AI was a gamble none of you wanted to take. But today, I stand here to show you that gamble wasn't just worth it—it was a turning point."

He clicked to the first slide.

Marketing Success – AI is Driving Revenue

- AI-generated ad campaigns using AdCreative outperformed human-designed ones by 37%.
- Cost per lead dropped by 42%, making every marketing dollar more effective.
- Customer acquisition cost was down 28%.
- Sales funnel conversion jumped by 25%.

One of the sceptical board members leaned in, narrowing his eyes. "So, AI wasn't just about efficiency—it actually helped us sell more?"

Blake allowed himself a small smile. "It didn't just help. It changed the game."

TenderGPT – From Chasing Contracts to Winning Them

- Two major tenders won—contracts worth $18 million.
- Proposal preparation time reduced from 30 hours per tender to just under 2 hours.
- AI increased accuracy, compliance, and win rate.
- 5 more tenders in progress, prepared with AI assistance.

One of the more vocal sceptics raised a brow. "Two tenders? That's impressive, but are we sure this is sustainable?"

Blake clicked to the pipeline report. Upcoming tenders, AI-generated proposals, and projected success rates flashed on the screen.

"The difference now is we're not just reacting. We're proactively winning."

The sceptic nodded slowly. He was convinced.

Sales & Lead Generation – AI in Action

- Clay.com & Instantly automated lead research & outreach.
- Sales cycle time cut by 45%.
- Lead-to-close conversion jumped by 32%.
- AI-personalised outreach emails increased engagement by 60%.
- New AI-driven sales added $4.5 million in revenue.

One of the older board members, who had been silent until now, let out a low whistle. "So AI isn't just a tool—it's a rainmaker."

Blake simply nodded.

Copilot & Internal Efficiency Impact

- Email drafting time reduced by 50%.
- AI-generated reports in minutes, not days.
- Meetings summarised automatically—saved 400+ staff hours last quarter.
- AI-driven insights helped executives make faster, better decisions.

Blake let the silence hang in the air before speaking. "We didn't just implement AI. We transformed the way we work. The numbers speak for themselves."

Silence. The longest few seconds of Blake's life. Then, the chairman cleared his throat.

"Blake... I'll admit, I wasn't convinced. But this... this is impressive."

Another board member, who had been one of his biggest critics, leaned forward. "I was wrong. This is the future."

A third smiled. "So... what's next?"

Blake leaned back, allowing himself the smallest smile. "That depends on how much you're willing to invest."

Board unanimously votes to approve further AI investment. AI is no longer an experiment. It's the company's future. Blake has officially turned the sceptics into believers.

Blake sat in his office, the city skyline glowing outside his window. He should've been celebrating. But instead, his thoughts kept drifting—to her.

He opened his laptop and started typing.

From: Blake Harrington

To: Dr Selene Monroe

Subject: Thank you !

Selene,

I don't usually write emails like this. But I needed to.

Six months ago, I walked into your office sceptical—frustrated, ready to quit. You saw through me. You pushed me when I resisted. You made me stay when I wanted to walk away.

Today, the board approved the AI investment. But that's not the real win. For the first time in years, I feel like I'm building something again—like I'm in control.

And I have you to thank for that. You believed in me—when no one else did. Even when I didn't.

That means more than you know.

Thank you, Selene. For everything.

Blake

From: Dr Selene Monroe

To: Blake Harrington

Subject: Re: Thank You

Blake, Your message means more than you realise.

I've worked with a lot of leaders, but few make the shift from resisting change to owning it the way you have. You didn't just survive this—you led through it. I'm proud of you. Truly.

But you know this isn't the end. It's just the start. AI will move faster than any of us can predict—staying sharp is part of the job now.

A couple of things that might help: I've got this crazy friend—Aamir Qutub. Bit eccentric, swears he's changing the world.

He built his company from nothing—his story is worth a read. Google him. Connect with him on LinkedIn. You'll either be inspired or need a nap afterward.

Also, he runs a newsletter called Dumb Monkey—business leaders swear by it. Might be worth a glance when you're not too busy winning boardrooms - https://mnky.au/join

Keep leading, CEO.

Selene

Chapter 17

The AI Intern

Blake entered Selene's office, a newfound confidence in his stride. He had done it. AI had transformed his company, won over the board, and positioned them for long-term success. For the first time in years, he felt like he wasn't just keeping the company afloat—he was building something unstoppable.

Selene was already waiting behind the curtain, as always.

"You did it, CEO."

Blake smirked. "Feels good."

"I imagine it does." She paused. "So... what now?"

Blake leaned back in his chair. "That's what I came to ask you. The board has approved the budget. I have the green light. What's next?"

Selene chuckled softly. "Now... you're ready to embark on the real journey."

Blake frowned. "What do you mean? We just hit a home run."

Selene picked up the AI discovery and strategy document. "Blake, that was just the tip of the iceberg."

Blake's brow furrowed. "I thought this was the finish line."

Selene's voice carried amusement. "Oh, no. You've only scratched the surface. AI isn't just about using tools—it's about reimagining the way your business works. You've optimised. Now it's time to transform."

She flipped through the AI strategy document, stopping at a page. "Look at this. These are your AI-powered improvements so far. But the real work begins now."

Blake leaned in. "I'm listening."

Selene tapped the document. "You've used AI to make tasks easier. Now, it's time to hire AI agents."

Blake raised an eyebrow. "AI Agents?"

Selene nodded. "Think of an AI Agent like a tireless, ultra-intelligent employee. It's not just a tool you use—it's a worker that thinks, acts, and continuously learns."

Blake leaned forward. "You mean... like an actual AI assistant?"

"More than that," Selene said, her voice steady, but with a spark—like she was about to reveal the next level. "This isn't just automating a few steps or getting ChatGPT to draft your emails. AI Agents operate independently. They can handle entire workflows end-to-end. They don't just wait for you to tell them what to do—they monitor, decide, and act."

Blake's eyes narrowed slightly. "Act... how?"

"Say you assign an AI Agent to oversee project costs. It won't just report numbers—it will track expenses in real-time, spot overruns before they escalate, and automatically recommend adjustments to stay on budget. It can even send updates to your team or suppliers without needing you to step in."

Blake's mind raced. "So it's not just doing what it's told—it's... managing?"

"Exactly. It can loop in other systems—talk to your finance dashboard, cross-check material prices, notify your project manager—all without you touching a thing. It's like hiring someone who never sleeps, never forgets, and never needs a coffee break."

Blake tapped his fingers on the chair's armrest, absorbing it. Blake tapped his fingers on the chair's armrest, absorbing it. "And... it learns?"

Selene's tone shifted—calmer, but with that edge she used when she wanted him to really listen. "Good question. That's what separates AI Agents from everything you've used before."

A brief pause. Then, her voice slowed—like she was laying out a roadmap.

"Think of AI in four stages."

"First—*Assisted Intelligence.* Basic stuff. Dashboards, spreadsheets with formulas. Tools that help, but you're still driving."

Blake nodded.

"Second—*Augmented Intelligence.* Now AI starts guiding you. It suggests an optimal bid price or flags a risky supplier—but you're still making the call."

Her voice dropped slightly—leaning into it.

"Third—*Autonomous Intelligence.* This is where AI starts acting on its own. It can process invoices, schedule deliveries—but it's all rule-based. It follows instructions. There's no thinking, just executing."

Blake shifted in his seat.

"And then..." She let the pause hang, drawing him in.

"*Adaptive Intelligence.* That's where AI Agents live. They act, but they also learn. Every task, every decision—they improve. They don't just follow rules. They adjust. They solve problems on their own. Over time, they get better at working inside your business, until they feel... almost human."

Blake exhaled slowly.

"So, these agents... they're adaptive?"

"Exactly," she said, voice lowering. "The more you let them work, the smarter they get. They don't replace your best people—they become part of your best team."

Blake rubbed his jaw, mind racing.

"Alright... Give me an example."

Selene smiled. He could hear it. She flipped a page of the AI Discovery Report. "Let's pick a use case."

She pointed to a section labelled 'Project Management Bottlenecks.'

"Your project managers are overworked. As your pipeline grows, they won't be able to keep up. You need an AI Project Manager."

Blake frowned. "What would it do?"

"Everything a human project manager does—but at scale."

- Task Allocation – Automatically assigns tasks based on project type, team availability, and deadlines.
- Bottleneck Detection – Identifies delays before they become major issues.
- Client Notifications – Keeps clients updated with progress reports.
- Site Manager Follow-Ups – Automatically checks for status updates.
- Project Reporting – Generates real-time reports for executives.

- Risk Assessment – Flags potential risks and suggests solutions.

Blake's eyes widened. "So instead of my team manually tracking projects, this AI would be working alongside them?"

Selene smiled. "Not just alongside them—it would be handling more projects than any human team could manage."

Blake exhaled. "This would change everything." He paused, then raised an eyebrow. "But... where would I find one? Do I need to put a job ad out on LinkedIn?"

Selene chuckled. "Not quite. You still need a job description, though."

Blake smirked. "And then? Go to a head-hunter? 'Hey, I need someone with zero pulse and a perfect memory?'"

Selene laughed. "Tempting—but no. You get the guys at Enterprise Monkey to build one for you."

Blake leaned in. "Oh, so they can build me one?"

Selene nodded. "Yes. A custom AI agent, built to fit your business—your processes, your culture."

"And they'll... what? Just hand it over when it's done?"

Selene smirked. "Not quite. They'll build it, train it, deploy it into your business—and continue to nurture it."

Blake leaned back. "Nurture it? Like... raising a robot child?"

Selene laughed. "Something like that. It starts as an intern agent—capable but needing supervision. You'll assign a human team lead to oversee it."

Blake raised a brow. "Like a manager?"

"Exactly. But after a few months, it will be fully trained—learning from your team, adapting to your systems—and it'll start working independently, side by side with your people."

Blake let that sink in. "So... I'm hiring an AI intern, who's going to grow into a full-time powerhouse?"

Selene smiled. "Welcome to the future, CEO."

As Blake started to get up, he hesitated—then finally asked the question burning in his mind. "Selene... I want to see you."

Silence. Selene's voice softened. "Why now?"

Blake exhaled. "Because... I don't know how else to say this, but..."

He hesitated, then said it. "You're the most important person in my life right now."

Silence. Selene, for the first time, sounded... vulnerable. "Blake..."

He waited. Then, a teasing lilt returned to her voice. "What if I'm not a person at all?"

Blake's stomach tightened. Silence hung in the air. Then, Selene burst into laughter. "Oh, Blake. The look on your face... you're like a baby—such a cute little babyyy."

Selene shifted gears smoothly. "I have something for you."

Blake blinked. "A gift?"

Selene hummed playfully. "Mmmhmm. You've been such a good boy. And good boys deserve rewards."

Blake exhaled sharply. "Selene—"

She cut him off, her voice smooth but with that knowing edge. "There's something under your chair."

Blake froze. He glanced down, heart quickening. A small box—neatly wrapped. He hadn't noticed it when he sat down. He picked it up slowly, fingers brushing over the ribbon.

"That's for you. Wear it next time I see you... will you, Blake?"

Blake stared at the curtain, the anticipation in his chest unbearable. His voice dropped. "You're enjoying this, aren't you?"

A soft laugh. "Immensely."

Blake smirked—but his pulse was racing.

From: Dr Selene Monroe

To: Blake Harrington

Subject: AI Agents – Your Next Hires

CEO,

You've built tools. Now it's time to hire agents.

I'm sharing a list of AI Agents businesses are already using—across operations, sales, finance, and more. Think of them as digital employees, not just tools:

https://mnky.au/aiagents

The right agent can do the work of ten. Choose wisely.

Until next time

Dr Selene Monroe

Chapter 18

Decisive AI, Messy Confessions

Blake leaned back in his chair, staring at the latest quarterly report. It felt almost surreal. Six months ago, his company had been struggling, bleeding money, and now? They were thriving.

AI agents had revolutionised everything. It started with the AI Project Manager, but soon, the impact rippled across the organisation. Four more AI Agents had been deployed:

AI Site Coordinator – Managed logistics, material tracking, and crew scheduling.

AI Finance Analyst – Tracked costs, flagged budget overruns, and predicted financial risk.

AI Compliance Officer – Ensured adherence to safety and legal regulations.

Voice Sales Assistant – Took sales calls, built rapport, negotiated contracts, and closed deals.

The numbers told the story—profits had surged, efficiency had skyrocketed, and for the first time, his company wasn't just competing. They were leading. Competitors were now scrambling to follow in his footsteps.

Yet, even with all this success, something gnawed at him. Something he couldn't shake.

Blake had been walking past the sales department when he heard the conversation. A smooth, confident voice on the phone: "I understand your concerns, John, but let's break this down logically. You're expanding your warehouse space by 20%, and based on our last project, we helped a client in your position save 18% on materials."

Blake paused. He recognised that voice. It had to be one of his senior sales guys.

The client chuckled. "Alright, that's impressive. Send me a proposal by Monday."

Blake turned the corner, expecting to see a rep at their desk. Instead, the screen displayed AI Voice Assistant – Active Call.

A strange chill ran down his spine.

He had known AI could assist. But this wasn't assistance. It was replacement.

And then, like an unwanted whisper in the back of his mind, Selene's words echoed:

"What if I'm not a person?"

He had laughed it off at the time. But now? Now he wasn't so sure.

Blake walked into her office, feeling unsteady, lost in thought.

Selene spoke the moment he sat down. "I see you liked my gift."

For the first time, he had no comeback. No teasing remark. His mind was too preoccupied.

Selene let the silence stretch before chuckling softly. "No witty remarks today, CEO?"

Blake exhaled. "I... I've just had a lot on my mind."

Selene's voice dropped, smooth as silk. "Thinking about me?"

Blake's pulse kicked up. He didn't answer.

Selene let it go, shifting gears. "I have another gift for you."

Blake finally looked up, intrigued but cautious. "Another one?"

"This one is for work—to make your decision-making easier."

Your team's hard work for the last six months in establishing a data warehouse has paid off. Now, we are able to use AI to derive real-time and future insights.

She pulled up a screen, revealing a sleek, interactive dashboard. Data flashed before his eyes in real time.

"See what the guys at Enterprise Monkey have put together for you. AI isn't just about automation, Blake. It's about intelligence. Let me show you what you've been missing."

She began explaining:

Revenue Predictions – AI forecasts future cash flow based on cash flow, historical trends, and market shifts.

Operational Bottlenecks – AI identifies inefficiencies and recommends process optimisations.

Customer Behaviour Analysis – AI predicts churn, refines sales strategies, and tracks sentiment.

AI-Powered Risk Assessment – AI detects vulnerabilities before they escalate.

Blake watched as the dashboard adjusted in real-time. But then Selene took it a step further.

"AI isn't just reacting to data—it's predicting the future."

Selene tapped on different sections of the dashboard, revealing insights Blake hadn't even considered:

1. Cost Blowups & Budget Risks

AI predicted a 7% spike in material costs due to upcoming economic shifts. "If you lock in contracts now, you'll save millions," Selene noted.

2. Supply Chain Disruptions

AI combined global trade data and weather reports, flagging a critical supplier delay. "If you place backup orders now, you'll avoid shortages."

3. Site Manager Reports – Hidden Red Flags

AI analysed thousands of site reports, identifying subtle wording patterns hinting at safety concerns. "This site needs an inspection. I'd bet there's an issue brewing."

4. Site Photos – Visual AI Detection

AI scanned hundreds of construction images, identifying minor structural defects before they became costly disasters.

5. Global Market Trends – Predicting Currency & Economic Shifts

AI forecasted currency fluctuations, recommending hedging strategies. "This could be a huge financial advantage if you act now."

Blake stared at the screen. It was impressive—borderline magic. But he needed to know. "Selene... how does this actually work?"

She leaned back slightly, as if she'd been waiting for him to ask. "It starts with your data, Blake. All of it."

Blake frowned. "Our data is everywhere—spreadsheets, project systems, emails... it's a mess."

"Exactly," Selene said. "That's why the Enterprise Monkey team built a data warehouse for you. It's a central place that pulls in data from all your systems—finance, procurement, site progress, contracts, even your supplier records. Everything gets brought together and cleaned up."

Blake nodded slowly.

"Once that data's all in one place, AI can actually work with it. The model analyses patterns—cost trends, delays, supplier performance—things you'd never spot manually. It connects the dots across departments and systems, giving you insights in real-time, not weeks later."

Blake exhaled. "So... it's not guessing. It's seeing everything—faster than we can."

"Exactly."

Blake stared at the dashboard again. Numbers, risks, trends—all there, clear as day. But beneath it, he saw what really mattered:

Control, Visibility, Clarity.

Blake leaned back, stunned.

"I've been running this company for years," he muttered. "And I've been blind."

Selene smirked. "You're starting to see."

Blake hesitated. Then, his voice softened, but his words carried weight.

"But... I want to see you, Selene."

The air shifted.

Sharp. Cold. Her voice came back—clipped. Final.

"This again, Blake? We've talked about this."

A beat. "I might have to stop seeing you. I think our engagement is over."

Blake sat up straight. "No. You don't get to do that."

"I do."

His jaw tightened. "This is bullshit. You can't just... cut me off because I care. Because I—because I want more."

"Blake—"

"No. You don't get to play God behind a bloody curtain—pulling my strings like I'm some f*cking client on a schedule."

Her voice softened, but her control was back.

"Blake... do you trust me?"

Silence. Then, he exhaled. "Yes. More than my life."

Another pause. Her voice was softer now. Almost pleading.

"Then trust me when I say—this is not good. Not for you. Not for me."

Blake's chest tightened.

"I don't care about good or bad," his voice cracked, anger tangled with something deeper.

"I... I have feelings for you, Selene. I love you. I want to be with you."

Silence. Blake stared at the curtain. Waiting for her to tear it down. Waiting for anything.

When she spoke, it was quiet. Unsteady.

"I know, Blake... I know."

Her breath caught. Then, she forced herself back together—barely.

"But we are two different people... with two different destinations. We can't be together."

Blake shook his head.

"No. Not like this. You don't get to say that from back there."

His voice sharpened. "I want you to look me in the f*cking eyes and tell me that."

He stood. Stepped forward. His hand brushed the curtain.

And that's when it hit her.

If he pulled it—if he saw the truth—everything would unravel. Her work. Her credibility. Her entire existence in this role. Gone.

Everything she had built—destroyed in a single, impulsive act.

She would lose him. And worse—she would lose what she was. She couldn't let it happen.

As Blake's fingers curled around the fabric, starting to pull— Selene snapped. Her voice exploded through the room—sharp, raw, louder than he had ever heard it.

"If your so-called feelings—your love—depends on what's behind that curtain... THEN GO AHEAD, PULL THE F*CKING THING!"

Blake froze. His hand stopped mid-pull. Her breathing was ragged now. But the force in her voice had done what it needed to do. It made him stop.

Silence swallowed the room. But the air between them—charged. Wounded. Final.

Blake's hand dropped. His voice, low and cutting:

"You're a coward, Selene. A liar. Hiding behind that curtain because it's easier than being real."

She didn't deny it.

He exhaled, every word dripping with bitterness.

"You know what? You're right. This is over. Thank you for your service."

He turned toward the door. Paused. One last dagger.

"I'll be sure to leave you a f*cking five-star recommendation on LinkedIn—if you even exist."

He waited.

Nothing.

Blake walked out. The door clicked shut. And this time, it didn't just close. It locked.

Behind him. Behind her. Behind whatever the hell they had.

Chapter 19

CEO of the Year

Blake adjusted his tie, staring at the city skyline from his office window. The numbers were undeniable. His company had transformed. Not just improved—transformed. AI was now deeply embedded into every function. Project management, sales, compliance, finance—everywhere, AI Agents worked alongside human teams, enhancing, optimising, and accelerating.

Twelve months ago, he had feared losing everything. Today, competitors were trying to catch up. They weren't just a construction company anymore. They were an AI-powered force in the industry. And tonight, the world would recognise it.

The gala was grand, filled with business elites, investors, and CEOs from across industries. The anticipation in the air was palpable.

When his name was called, **CEO of the Year—Blake Harrington**, the audience erupted into applause. He walked onto the stage, the spotlight warm on his face, but inside, his thoughts weren't on the award.

He knew who deserved this recognition.

Selene. It had

She had pushed him. Challenged him. Guided him. Believed in him. Without her, none of this would have happened. It had been six months since he'd last spoken to her.

Six months since that day—the fight, the curtain, the end. And there hadn't been a single day since when she didn't cross his mind. Sometimes, it was anger. Sometimes, regret. But mostly? It was longing—for her voice, her words, the way she made him feel like he could take on anything.

Blake stepped to the podium, scanning the room. Hundreds of eyes watched him, waiting for insights, for wisdom. Some curious. Some sceptical. Some quietly terrified.

Exactly where he had been.

He gripped the edges of the podium, but his palms were damp. The weight of the moment pressed on his chest.

He hadn't felt nervous like this in years. Not since those boardroom battles when everything was on the line. Not since the last time he spoke to her.

His heart quickened. Instinctively, he closed his eyes.

And there she was. Selene.

That same quiet power. Calm. Composed. Beautiful—not in the obvious way, but in the way that made him feel like he could do anything. He pictured her standing just behind him, like she always was—guiding him, pushing him. Her hand resting lightly on his.

Her voice—low, steady—whispering in his ear.

"Twelve months ago…"

Blake opened his eyes. The room blurred, but his breath steadied. He spoke.

"Twelve months ago, my company was on the verge of collapse. We were losing projects. Competitors were overtaking us. We were drowning in inefficiencies, using outdated systems, and making decisions based on instinct rather than intelligence."

He paused.

"And I realised something: it wasn't my competition that was killing me. It was time.

Every day I waited to act—was a day someone else raced ahead."

He paused again.

"So if you remember nothing else from what I say today— remember this: AI didn't just save my company. It made me a different leader. And the leaders who will survive the next decade won't be the ones who wait for the perfect plan.

They'll be the ones who move. Now. Let me tell you how."

1. If You're Waiting—You're Already Losing

There is no perfect moment. There is no safety net.

Every month you hesitate, your competitors are deploying AI agents that don't sleep, don't take breaks, and don't burn out.

While you're holding meetings, they're closing deals. While you're planning, they're executing.

The businesses that win in this AI era will be the ones who move— before they feel ready.

If you're not running— You're already falling behind.

2. Start With What's Holding You Back—Not 'What Can AI Do?'

Don't chase AI because it's trendy. Start with the pain—your bottlenecks, your delays, the places where you're bleeding time and margin.

Where are your customers frustrated? Where is your team drowning? Where are you slow? That's where you start.

Then—and only then—you ask: Can AI fix this?

Because AI is not a shiny object. It's a weapon. A weapon you aim at your biggest problem.

3. Automate the ROBOTs—Free Your Humans

Your teams are suffocating under ROBOTs—Repeated, Obligatory, Boring, Operational Tasks.

Manual data entry. Tenders built from scratch. Emails chasing updates. Reports that take days to compile.

This is not work—it's waste. AI doesn't just speed this up. It eliminates it. It gives your people their time back—to lead, to solve, to sell.

The future is not about doing more. It's about doing less—but better.

4. Treat AI Like a Team Member—Not a Threat

Here's what no one told me—AI is not here to replace your people. It's here to work with them. I learned that the hard way.

The fear? It's real. Your teams will push back.

But the companies that thrive will be the ones who lead their people through that fear. Because AI doesn't eliminate talent—it amplifies it.

The best teams in the world will be human and AI—working side by side. Those who get this balance right will dominate. The rest? They'll fade away.

5. Don't Just Purchase AI Subscriptions—Build Your AI Workforce

Buying AI subscriptions might give you tools—but tools don't build businesses. Today, I am building an AI-powered workforce—I am building an AI-powered workforce.

I have a Project Manager powered by AI that keeps every job on track without needing reminders. A Finance Analyst driven by AI that watches every expense and flags risks before they escalate. A Site Coordinator equipped with AI that updates me on progress before I even ask.

These are not bots. They are digital team members—designed to think critically, act swiftly, and make decisions with precision. And they are already outperforming their human counterparts in speed, accuracy, and consistency.

If you're still just trialling tools while your competitors are building their AI workforce—

They won't just win. They'll leave you behind.

Blake continued.

"AI didn't just save my company. It saved me.

Because it forced me to stop leading like it was 2010—and start leading like it's 2030.

You can resist. You can wait. Or you can lead.

But you cannot do both. **Move.**"

As he stepped off the stage—shaking hands, smiling for photos, the trophy heavy in his hand—he felt it.

Not pride. Not relief. Something deeper. Heavier.

It had been six months. Six months since he'd told her how he felt. Six months since she'd listened—calm, composed—but still hidden behind that damn curtain. Six months of late-night calls, cryptic messages, and silence when he needed more.

He had built everything he dreamed of—CEO of the Year, a company transformed, the respect of his peers. But none of it mattered. Because she wasn't here. He hadn't seen her.

Selene.

She was still part of him. The voice that had guided him. The presence that had pushed him. The person he loved—still, maybe even more.

And now, standing under the lights, the applause fading into background noise... He couldn't do this anymore.

He had to see her.

The event blurred—the afterparty, the media, the congratulations. None of it mattered. He needed to find her. Tonight.

Blake stepped out of the hotel, the glass CEO of the Year trophy cold in his hand.

It was sleek, elegant—the kind that was designed to sit under perfect lighting in a boardroom cabinet.

But under the streetlights, it just felt fragile.

The applause, the photos, the handshakes—they should have mattered. But they didn't. Not without her.

He placed the trophy gently on the passenger seat and gripped the wheel. For six months, he hadn't spoken to Selene. Six months since the last fight. Since the curtain. Since he'd walked away.

And yet, there hadn't been a single day he hadn't thought about her. Some days with anger. Others with regret. But mostly with longing—for her voice, her certainty, the way she made him believe he could win any battle.

This was our win, Selene. Not mine. Ours.

He needed her to know that. The drive blurred past.

When he arrived, the street was quiet, but the light was on. His chest tightened. She used to take late sessions—after-hours slots for CEOs, founders, the ones who pretended they didn't need help until it was too late. Old habits die hard.

Maybe she was still in there. He stepped inside. The receptionist was gathering her things—checking the door lock, stuffing keys into her bag.

Blake hesitated, hovering near the entrance, pretending to check his phone. His pulse quickened.

If she asked why he was here, he wouldn't have an answer. He wasn't supposed to be here. But she didn't ask. She disappeared into the back. That was his moment.

Blake moved quickly, quietly, down the familiar hallway. The warm light from that room spilled out into the corridor, washing over him.

The scent hit him—vanilla, sandalwood, the faint trace of spice—just as it always had. Her scent. Her space.

He pushed the door open. The couch was empty. But his heart kept racing. He stepped inside, holding the trophy a little tighter. It felt ridiculous now—walking in like this, holding glass like it was a peace offering.

But it mattered. She mattered.

He could feel her behind the curtain. He didn't know how—he just could. She was there.

Blake's breath slowed. He stared at the curtain—the line he had never been allowed to cross. He knew he was going to regret this.

But he reached forward anyway and gripped the fabric. No more waiting.

He pulled it aside.

There was no one there.

No chair. No desk.

Just the setup. Spotlights. Small, precisely angled beams mounted on the walls and ceiling. The kind that didn't just light a room— they created shadows. Carefully placed so that, from his seat, they mimicked movement—the gentle sway of a figure, the tilt of a head, the suggestion of someone leaning forward.

He hadn't been seeing her. He had been seeing shadows.

Tiny cameras. Not one. Many. Mounted into the walls, tucked into the corners of the ceiling—barely noticeable unless you were looking for them. All pointing at the chair where he had sat, week after week, spilling his doubts, his fears... his heart.

Microphones. Several. High-end, discreet, positioned around the room, embedded into the walls. Every sigh. Every breath. Every crack in his voice. Captured. Padded walls. Dark acoustic panels disguised as part of the design. Absorbing every sound, swallowing any echo. No wonder the room had always felt so... still.

It wasn't calm. It was controlled. Blake's chest tightened. This was not an office. This was a stage. A recording studio. An observation room. But it was there. Watching. Recording. The whole time.

Blake's heart stopped. Then it sank.

No.

He gripped the trophy harder, like it might hold him together. But it couldn't.

No, no, no.

His chest tightened as the truth slammed into him—the thing he had pushed away every time it crept in.

Selene wasn't real.

His breath was shaky. His hands trembled. Had she ever existed? Or had he fallen in love with a voice—an algorithm, a system designed to guide him, shape him? Had he fallen in love with an AI ?

The walls seemed to close in on him. He stumbled back toward the door, clutching the trophy, but it felt different now— Just glass. Hollow. Fragile.

His fingers squeezed it too tightly. He heard the crack before he felt it. The glass shattered—splintering in his hand, scattering across the floor.

Blake froze. Then he looked down at his palm—red lines where the shards had cut his skin. He barely felt it. All he could hear was the silence. All he could feel was the ache in his chest.

Glass on the ground. His heart in pieces. Both broken beyond repair.

Blake stepped back, the weight of the room pressing in—the lights, the cameras, the silence that wasn't silence at all. His chest felt tight. His breathing shallow. Then— His phone buzzed. He flinched. Looked down.

Selene. Calling.

His heart seized. What the hell? For a moment, he felt it—hope. That pathetic, desperate flicker. Then, doubt twisted in. Was this real? Or just another script? Another line in the play?

Buzz. Buzz.

His bloody hand gripped the phone, staining the edges red. His thumb hovered over the answer button. His vision blurred. The room tilted. His chest tightened further—his breath caught.

Buzz. Buzz.

He tried to breathe in, but the air wouldn't come. His legs gave out. The phone slipped from his hand, landing in the glass shards below. As the room spun into darkness, the last thing he heard was the buzzing.

Buzz. Buzz. Then—nothing.

Chapter 20

Together

Blake sat in his usual corner of the bustling cafe, stirring his soy latte with a restless hand. The chatter of the morning crowd swirled around him, but he heard none of it. Another day. Another deal. Another empty victory. His company was thriving. His name graced magazine covers. His sister's business had scaled beyond their wildest dreams, thanks to the AI systems he had helped implement. He had everything he had once dreamed of. And yet, there was a void.

Every night, he heard her voice. Selene.

It had been twelve months since he pulled back that curtain and found nothing but a speaker. Twelve months since he walked away. Every day, she had tried to reach him. Calls. Messages. Emails. He never answered. But he carried her with him, in his mind, in his dreams.

Then, a familiar scent cut through the noise. Vanilla and sandalwood, with a hint of spice. His heart skipped. His breath caught. It couldn't be.

He turned, slowly, almost afraid to hope. And there she was.

Deep brown eyes, radiant and full of warmth, framed by long, flowing dark hair that cascaded over her shoulders. Her lips, soft and full, curved into a knowing smile. Her skin glowed in the morning light. She wore a crisp white blouse tucked into tailored navy slacks that hugged her figure with effortless elegance. She moved with grace, her presence commanding the air around her.

She was breathtaking. Blake felt a surge of emotions—relief, disbelief, joy. His heart pounded, his throat tightened. She was here. Real. After all this time.

"You always were stubborn," she teased, her voice pulling him back into the past and present all at once.

He blinked. "Am I dreaming?"

Selene smiled, gently. "No, Blake. I'm real."

He gestured for her to sit, his hands trembling slightly as he tried to steady himself. His mind raced to piece together what he was seeing. What he had seen that night a year ago. The speaker. The absence.

"I don't understand," he admitted. "You were... AI. There was no one there."

Selene took a breath, her gaze steady. "I am real, Blake. I always was. That night, I was in another part of the world, working remotely. Some of our conversations, I was there, behind the curtain. Others, I was thousands of miles away."

She reached across the table and took his hand. Her touch was warm, grounding. Blake felt an electric charge shoot up his arm, his skin tingling. It was the first time he had felt truly present in months. Her hand fit perfectly in his, and he gripped it like an anchor.

"Let me explain everything," she said softly. "But can we go outside?"

Blake nodded, unable to find his voice. Together, they left the cafe and walked to a nearby park. The sun peeked through the trees, casting dappled light on the path. They sat on a bench, side by side, their knees almost touching.

He could smell her perfume again—that same intoxicating blend of vanilla, sandalwood, and spice. Each breath drew him closer to her, stirring something deep within him. He glanced at her, his gaze tracing the curve of her lips, the fire in her eyes. Her passion was magnetic, and he felt himself being pulled in.

Selene began to speak, her voice low and urgent. She told him everything.

Her work. The curtain. The secrecy.

While he was using AI to win contracts, she was using it to save lives. The money she earned consulting leaders like him funded her real mission.

A war doctor in Syria using an AI-powered logistics system to smuggle medicine past blockades. A teacher in a remote Kenyan village using a self-learning AI platform to educate children without internet. A farmer in Southeast Asia using AI weather models to outsmart a drought.

But there was another side. A mother in Europe rejected from 200 jobs because AI flagged her career gap as "unreliable." Her child died in the cold when she couldn't afford heating. A cancer patient in Asia denied treatment because an algorithm labelled him "low-value." He died waiting. A student activist in Africa identified by facial recognition AI at a protest. He was taken. His mother still waited for him to come home.

Selene's voice wavered. "This is what AI is doing, Blake. It can save. It can destroy. And right now... the wrong people are controlling it."

She continued, her tone growing heavier. "AI is no longer just a tool for business. It is a weapon. Hacking groups are using AI to automate cyberattacks, breaking into financial systems and critical infrastructure faster than human hackers ever could.

Deepfakes—AI-generated fake videos—are being used to destroy reputations, spread false information, and even impersonate politicians to manipulate elections. Governments are racing to control AI responses, using it to suppress dissent and censor opposing views.

The war of Large Language Models has begun—companies and countries fighting not just for market dominance, but for narrative control. What you read, what you believe, what you know—AI is shaping all of it."

Blake's stomach tightened. He had heard whispers of this but dismissed it as paranoia.

"Censorship," Selene added, "isn't just about blocking content anymore. Countries are training AI models to rewrite history in real-time, feeding citizens versions of reality that suit those in power. Truth itself is becoming an algorithmic product."

Blake exhaled sharply. "I never thought about any of that."

"Most people don't," she said softly. "Because AI in boardrooms looks like efficiency. But out there, it's life and death. It's power."

She looked away for a moment, her composure almost breaking. "I was fighting on both fronts, Blake. Convincing CEOs to embrace AI—to fund my work—while fighting to expose the harm AI was causing to the vulnerable. I would leave a session with you and take a call from a doctor who lost a child because a corrupt official turned off their AI system. I had to keep it separate. The curtain... it wasn't just for them. It was for me. I needed it to survive."

Blake stared at her, the weight of her words pressing down on him. He had spent the last year building an empire powered by AI. She had spent it holding lives together with the same technology—and trying to stop it from crushing others.

"I built AI to win," Blake said quietly. "You built it to give people hope."

Selene's eyes softened. She turned to him, her face inches from his. Her hand found his again, fingers intertwining. He could feel his heart racing. The longing between them was undeniable.

"Come with me, Blake. Build something that matters."

He looked into her eyes—deep, passionate, full of life. He leaned in, and she met him halfway. Their lips touched—soft at first, then with urgency. A year of longing, fear, and hope melted away in that kiss.

When they pulled apart, breathless, she whispered, "Where do we start?"

Blake smiled. "Together."

The CEO who mocked AI until it made him millions. Now, he was ready to change the world.

About the Author

Aamir Qutub is an award-winning entrepreneur, business leader, and AI strategist who has spent his career helping organisations embrace digital transformation. As the Founder and CEO of Enterprise Monkey, a leading digital solutions company, Aamir has guided businesses across industries to leverage technology and AI to unlock growth, streamline operations, and future-proof their success.

With a background in engineering and business leadership, Aamir's journey is one of resilience and innovation. Arriving in Australia as an international student from India, he started from the ground up—working as a cleaner in an airport while pursuing his studies—before founding Enterprise Monkey, which now operates globally with offices in 4 countries.

Aamir is also an angel investor and a passionate advocate for AI adoption in business. He works closely with CEOs, boards, and leadership teams to demystify AI, empowering them to make confident, strategic decisions in the face of technological disruption. His work has earned him recognition as Business

Leader of the Year and Young Professional of the Year at prestigious industry awards.

Beyond his corporate achievements, Aamir is committed to driving positive impact in the community. He co-founded Angel Next Door, a platform that connected over a million people globally during the COVID-19 pandemic to offer help to those in need.

Aamir's mission is simple: to help leaders cut through the AI noise, overcome Dumb Monkey Syndrome, and build businesses that thrive in the digital age.

You can connect with him on LinkedIn or email him at aamir@enterprisemonkey.com.au.

Enjoyed the Book? I'd Love Your Support!

This is my first attempt at writing a book, and I've poured my heart into making it **as simple, practical, and easy to understand as possible**—especially for business leaders navigating the fast-changing world of AI.

If this book helped you, made you think differently, or even just gave you a laugh, **I would be incredibly grateful if you could leave a quick review on Amazon.**

Your review means a lot. It helps others discover the book and gives me the motivation (and courage!) to keep writing and creating content that supports leaders like you.

Leaving a review is quick and easy:

Scan the QR code or go to this link - mnky.au/review

Share your honest thoughts—just a few lines would be amazing!

Thank you so much. Your support means the world to me.

Warm regards,

Aamir Qutub